THEY STAND ON GUARD

A DEFENCE DIRECTION FOR CANADA

Copyright © 1991 by Conference of Defence Associations

Published by the
Conference of Defence Associations Institute
601-100 Gloucester Street
Ottawa, Ontario
K2P 0A4

THE VIEWS EXPRESSED IN THIS BOOK ARE THOSE OF THE
AUTHOR AND DO NOT NECESSARILY REFLECT THE
OFFICIAL POLICY OF THE CONFERENCE OF DEFENCE
ASSOCIATIONS

Cover: Canadian Forces
ISBN 0-921687-08-7

Printed in Canada by
Tyrell Press Limited, Ottawa

THE CONFERENCE OF DEFENCE ASSOCIATIONS INSTITUTE
AND
THE UNITED SERVICES INSTITUTE OF OTTAWA

Publication of this book was made possible by the Conference of Defence Associations Institute with generous assistance from the United Services Institute of Ottawa.

The CDA Institute publishes books, policy papers, a magazine and a newsletter. It also sponsors conferences and seminars, and produces defence-related film projects — all with the aim of placing before the people of Canada the problems of defence and the complex issues involved. Such activities contribute to an informed exchange of ideas on defence and will result, it is hoped, in a more positive attitude towards the need for improving the capability and efficiency of our defence forces.

The United Services Institute of Ottawa is a group of loyal Canadians who are concerned with the military, technical and economic stability of Canada and who want the nation to grow and fulfill its responsibilities within the world community. It also supports the need for strengthening and properly equipping our Armed Forces in order to preserve our sovereignty and national security.

Views and recommendations put forward in this book are those of the author, Bob Hicks MP. They do not necessarily reflect the views or the policies of either the CDA Institute or the United Services Institute. Both organizations congratulate Mr. Hicks on his initiative in writing about testimony to the National Defence and Veterans Affairs Committee and his courage in making his personal views public. We believe that the Federal Government should be more open and receptive to public debate on defence issues. Books such as this provide background information which will assist the electorate when this more enlightened policy is finally adopted.

This book is dedicated to all those members of the Canadian Armed Forces, past and present, who have for so long done so much with so little. BZ

ACKNOWLEDGEMENTS

Many people have participated in the production of this book, but I would especially like to recognize the exceptional contribution of David E. Code and R. Bruce Wallace, without whose tireless attention to research and obvious love for the subject this book would not have been possible.

I also wish to extend thanks to Sharon Hobson for her fine research for Chapter VI, on industrial support for defence.

Countless conversations have been held with many others and I am grateful for their contributions. In particular I wish to thank Major General Dan G. Loomis (ret'd), Brigadier General Keith R. Greenaway (ret'd), W.G. Hillaby for his assistance with computer editing, and certain naval personnel who were too modest to be named.

About the Author

BOB HICKS

Was an educator in the City of Scarborough for 30 years before being elected to the House of Commons.

After being elected in 1984 and re-elected in 1988, has dedicated nearly all of his time to defence related activities.

He has been a member of the Standing Committee on National Defence and Veterans Affairs for nearly 7 years.

Has been Chairman of the Canadian NATO Parliamentary Association for 4 1/2 years, and is therefore the leader of the Canadian delegation for meetings of the North Atlantic Assembly held in all 16 NATO Nations.

Was elected Vice-President of the North Atlantic Assembly (the Parliamentary wing of the 16 NATO Nations) in November 1990.

Has been Chairman of the P.C. Caucus Committee on National Defence since 1986.

GLOSSARY OF ACRONYMS AND ABBREVIATIONS USED IN THIS BOOK

AMF(A)	Allied Command Europe Mobile Force (Air)
AMF(L)	Allied Command Europe Mobile Force (Land)
AWACS	Airborne Warning and Control System
CAST	Canadian Air-Sea Transportable (Brigade)
CD	Conference on Disarmament (UN)
CF	Canadian Forces
CFE	Canadian Forces in Europe
CFE	Conventional Forces in Europe (conference)
CISS	Canadian Institute of Stategic Studies
CSCE	Conference for Security and Cooperation in Europe
CUSRP	G Canada-US Regional Planning Group
DEA	Department of External Affairs
DFO	Department of Fisheries and Oceans
DIPAC	Defence Industrial Preparedness Advisory Committee
DIPP	Defence Industrial Productivity Programme
DND	Department of National Defence
IISS	International Institute for Strategic Studies
DPSA	Defence Production Sharing Arrangements
LRPA	Long Range Patrol Aircraft
MOT	Ministry of Transport
NATO	North Atlantic Treaty Organization
NORAD	North American Aerospace Defence
PJBD	Permanent Joint Board on Defence (Can-USA)
RCMP	Royal Canadian Mounted Police
SAR	Search and Rescue
SCOND	Standing Committee on National Defence
SCONDVA	Standing Committee on Defence and Veterans Affairs
SSN	Nuclear powered submarine
TRUMP	Tribal Class Update and Modernization Program
UN	United Nations organization
WTO	Warsaw Treaty Organization (Warsaw Pact)

TABLE OF CONTENTS

I INTRODUCTION   **1**
   A. Of Policies and Roles   1
   B. The Commitment-Capability Gap   7

II DEFENDING CANADA   **13**
   A. Canada's Interests   13
   B. Does Canada Need Defence?   15
   C. Is Defence Moral?   16
   D. The Changing International Scene   18
   E. Do We Need Alliances?   21
       1. Lessons learned   21
       2. Accepting Responsibility   23
       3. NATO - An Effective Alliance   24

III CONSIDERING THE OPTIONS   **29**
   A. The Present Situation   29
   B. The Pillars of Defence Policy   37
       1. Nuclear Deterrence   37
       2. Conventional Deterrence and Defence   41
       3. Sovereignty   44
       4. Peacekeeping   48
       5. Arms Control and Disarmament   53

IV FUTURE ROLES, COMMITMENTS AND EQUIPMENT   **63**
   A. The Constants   63
   B. The Variables   64
   C. Considerations and Priorities for Defence   65
       1. Maritime Forces   69
       2. Land Forces   82
       3. Air Forces   91
       4. Reserves and Cadets   102

V LEADERSHIP ISSUES — **111**
  A. Policy — 111
  B. Social Issues — 112
  C. Budget — 117
  D. Public Support — 120

VI INDUSTRIAL SUPPORT — **125**
  A. The Essential Foundation — 125
  B. Major Industrial Sectors — 125
    1. Aerospace industry — 126
    2. Defence Electronics — 127
    3. Shipbuilding and Ship Repair — 128
    4. Automobiles — 130
    5. Munitions — 131
  C. The Market and Government Support — 132

VII SUMMARY OF RECOMMENDATIONS — **141**

I INTRODUCTION

A. Of Policies and Roles

We Canadians like to think of ourselves as a nation of clean-living, peaceable people, uninvolved in international machinations and powerplays, somehow above the sordid little games of territorial grabs and influence peddling or of trying to change the governments of other sovereign states, yet always ready and willing to do our share for world peace. Indeed, we have never started a war on anyone, never fought a war alone against anyone, never displayed any territorial ambitions. Moreover, we were among the first to sign up for peacekeeping operations under the United Nations when this was proposed by our own Secretary of State for External Affairs, L.B. Pearson, in 1956.

Of all the roles and activities of the Canadian Armed Forces none is more universally accepted than peacekeeping. Even the most vociferous critics of Canada's security and defence policies concede that the intervention of Canadian peacekeeping troops plays a valuable role in keeping combatants apart and preserving a semblance of peace. Because peacekeeping is seen as non-violent, and helpful, Canadians are united in calling it an acceptable, even commendable, use of military resources.

But we Canadians must also ask ourselves whether it is possible to base an entire defence policy on a peacekeeping role alone. The events in the Persian Gulf area in 1990 and 1991 offer a sharp reminder to us that our association with the United Nations does not allow us to hide from the reality of military force, cannot disguise the uncomfortable truth that military force may be needed occasionally in the cause of restoring peace and justice to victims of aggression.

What then is Canada's defence policy? and what are the roles of the Armed Forces?

Our defence policies appear to be sound enough. As enunciated from time to time by Ministers and other representatives of the Department of National Defence (DND) they are more or less this:

we stand for peace, freedom, human rights and the dignity of the person; and our defence capability exists not so much for the defence of Canadian territory as for the defence of our values and our hard-won rights and freedoms.

Defence and External Affairs documents also remind us that Canada's security policy rests on three major components:

- defence and collective security,
- arms control and disarmament,
- the peaceful resolution of disputes.

The Department of National Defence and the Canadian Forces support this security policy by their contributions to:

- strategic deterrence
- conventional defence
- sovereignty
- peacekeeping
- arms control

These elements of policy have been with us for many years, and so have the stated roles of the Forces. So it is worth while, during the current discussions, to consider some of the principal statements on the subject that have appeared over the past couple of decades. We might begin with the Defence Structure Review of 1975, which listed four roles for the Canadian Forces, as follows:

1. The protection of Canada and Canadian national interests at home, i.e. sovereignty,

2. The defence of North America in cooperation with US forces,

3. The fulfillment of NATO commitments as may be agreed upon,

4. International peacekeeping duties.

For the roles of Maritime Command, we can look to the Defence White Paper statement of 1987, which declared: "Canada is a maritime nation with a proud sea-going tradition. The three oceans off our shores are sources of natural wealth, which we are only beginning to tap...Canadian naval forces must be able to respond to challenges within our own waters, if necessary denying their use to an enemy." In the words of a former Deputy

Commander of Maritime Command "our country is a maritime state, necessitating a three-ocean policy with strategic, economic, legal and sovereignty dimensions. It formulates the strategic concept of sea control in the three oceanic approaches to this country. (but)...To date the Department of National Defence has not been sufficiently precise in defining the navy's roles and missions in our three oceans..."[1]

The statement of objectives of Mobile Command (Army) states its responsibility for surveillance and control and security within the national territory of Canada, and adds the responsibility for "provision of resources for United Nations Peacekeeping operations".

The objectives for Air Command charge it with surveillance and control over Canadian airspace, and adds responsibility for air support for Maritime and Land Forces in Canada, airlift requirements worldwide of the Canadian Forces, and coordination and conduct of search and rescue operations.

The National Defence Estimates for 1989-90 contain statements of objectives for each of the Maritime, Land and Air Forces, but they all bear a remarkable resemblance. All are to

> *"maintain combat-ready general purpose (maritime/land/air) forces to meet Canada's defence commitments, with emphasis on surveillance and control of Canadian territorial waters, land and air spaces as appropriate ...and to provide assistance to civil authorities in the event of emergency or disaster."* [2]

The need for Canadian Forces was defined for several decades by the Cold War. Defence policies for tomorrow must reflect the dramatic changes that began in 1989-90 in East-West confrontation and in the Soviet Union, including the equally dramatic changes caused by the demise of the Warsaw Pact and the decline of Communism in Eastern Europe, the formation of a united Germany, and the conflict over resources vital to Western prosperity, such as oil in the Middle East and the UN war for the liberation of Kuwait. Threats to international peace and the security of Canada may change uncontrollably in the future, but it is clear that the need for Canadian military forces, described in Chapters III and IV, will remain.

---

[1] F.W. Crickard, RAdm (ret) "The US Maritime Strategy - Should Canada be Concerned?" in "In Defence of Canada's Oceans" Conference of Defence Associations, 1988.

[2] National Defence 1989-90 Estimates, pp. 28, 25, 41.

The needs as well as the objectives of the Canadian Forces were well expressed in "Challenge and Commitment, A Defence Policy For Canada" tabled in Parliament on the 5th of June, 1987 by the Hon. Perrin Beatty, the Minister of National Defence. It was the first defence white paper in 18 years, and it was a very important document - as much for its tone as for its content.

The Minister could have glossed over the problems of Canadian defence, could have given Canadians the "feel-good" treatment. Instead he courageously presented the facts, warts and all, about the concerns and shortcomings of Canada's defence capabilities and the challenges for which it was responsible. He set out his and the department's assessment of the international environment, and the nature and extent of the military threat, and pointed the way ahead for Canada's Armed Forces.

Among the many features of the White Paper were the announcements of:

- Total Force concept, including the integration of the Regulars with the Reserves, which would rise gradually to 90,000 from the then 23,000

- Acquisition of 10 - 12 nuclear - powered submarines in order to provide improved 3-ocean patrol capability

- Consolidation of land force commitments in Europe to the Central Front "to provide a credible and more sustainable Canadian contribution" and to be a more honest and responsible partner to our allies

The language in this historic White Paper carried the tone of the Cold War; but that was honest in 1987, since the Cold War was still very much on. This tone has since become dated, in view of the rapidly changing events in Europe and on the disarmament scene. Before the year was out the INF treaty for reduction and elimination of the longer range intermediate nuclear missiles had been agreed upon. Soon after, new life was breathed into the strategic missile negotiations at Geneva; and the CFE talks (Conventional Forces in Europe) were begun at Vienna. The CFE resulted in an arms control treaty signed in Paris on November 19,

1990 by all members of the two alliances, and by the end of the year the Soviets had begun removing their forces and their equipment from the occupied countries of Central Europe.

Frankly, I am glad that the Cold War aspects of the White Paper are out of date now, glad that Canadian and NATO policies have succeeded in reducing the danger of war in Europe. Nevertheless, there are many uncertainties ahead of us.

Considering the conditions that existed prior to June of 1987, the White Paper was very sound. Why then did some Canadians react adversely to it? One of the reasons was that this was the first time in their memories that they had seen the facts of sovereignty and international confrontation spelled out so clearly. In 1987 there was no pussy-footing, as there had been in the past. There were also a few unabashed references to the military threat - a consideration that many Canadians had long preferred to ignore. To some, the term "threat" sounds bellicose, when merely it is the defence strategist's way of referring to the potential military challenge.

The 1987 White Paper contains much that is important for Canadians, much that is still worth remembering:

"The first objective of Canada's security policy is to promote a stronger and more stable international environment in which our values and interests can flourish." [3]

"Canada has no aggressive intentions toward any country." [4]

The White Paper also gave life and meaning to the expression of Canadian sovereignty, which had been practically ignored since being highlighted over a decade before. It is unfortunate that so much attention was allowed to focus on the issue of the nuclear-powered submarines, and unfortunate that the focus was allowed to be so wrong. The government allowed a noisy opposition to repeat, almost unopposed, the false accusation that these SSNs were to be nuclear ARMED, when in fact they were to be nuclear POWERED. These opponents were also allowed to concentrate their fire on the celebrated $8 billion cost of the program, without a clear understanding that this cost would be spread out over the 27 years of the acquisition schedule.

[3] Challenge and Commitment, page 3.
[4] Ibid, p 17

The cost was high but, put in annual terms, it was no larger than other major acquisition programs. DND considered it was excellent value for the money, and competent authorities in other countries agreed (with the exception of the US Navy, but they had their own reasons for disliking our nerve in daring to defend and perform surveillance of our territorial waters).

The prime benefits which would have come from this new class of vessel would have been not only in the Arctic, where they will remain the only truly capable vessel, but also in the two more traditional Canadian oceans - the Atlantic and the Pacific. The SSN is a vessel of manoeuvre, not merely one to be positioned in the path of a possible intruder. With its speed and endurance, the SSN would have enabled Canada's armed forces to patrol large expanses of water with fewer vessels and greater efficiency. As one submarine commander said:" For three months you don't have to feed it, fix it, fuel it or speak to it - the SSN is the complete lone ranger."[5]

Oberon class, conventionally-powered submarine.

[5]"The Case for Nuclear Submarines" by John Dyson, *Readers' Digest*, November 1988.

The government must bear the responsibility for the failure of this program and the damage done to the White Paper. Taking refuge in silence, they allowed the impression to remain that the various charges were correct. But the charges were false. Table 1 offers a useful cost comparison of figures drawn from the daily press on different occasions during 1988.

Table 1.

To Pay For	ANNUAL COST ($ Billions)	27 YEAR COST ($ Billions)
SSN SUBMARINES	0.3	8.0
TOTAL CAN DEFENCE	11.2	302.0
TOTAL SOCIAL PROGRAMS	63.0	1,701.0
SERVICE NAT'L DEBT	32.0	864.0
UNEMPLOYMENT INSURANCE	10.6	286.0
OLD AGE SECURITY	14.6	394.0
FAMILY ALLOWANCE	2.6	70.0
ILLEGAL DRUG USE	11.0	297.0

B. The Commitment-Capability Gap

In his introduction to the 1987 defence White Paper Prime Minister Mulroney said:

> *No government has a more important obligation than to protect the life and well being of its people...Canada must look to itself to safeguard its sovereignty and pursue its own interests. Only we as a nation should decide what must be done to protect our shores, our waters and our airspace.*

On the facing page the Minister, Hon Perrin Beatty, spoke of the need for honesty about our defence capabilities, of the need for a manageable mandate, the resources to do the job and the support of the Canadian people.

The White Paper then listed a number of uncomfortable facts concerning Canada's defence commitments and its capabilities, in these terms:

- after years of neglect our equipment is obsolete,
- we have allowed our contribution to equipment purchases to fall as low as 9% of budget in some years, compared to the NATO average of 25%
- the navy has too few vessels, no minesweepers to keep our harbours clear, and the newest ship was then 14 years old (make that 18 years as of today)
- the militia is too small, and badly equipped
- there are too few maritime patrol aircraft
- there is no provision for lost CF-18 aircraft
- logistic and medical support for our forces committed to Europe is lacking
- the root of the problem is the level of funding available to defence over the last 25 years (which has been allowed to slide from more than 20% of federal spending to less than 10%)

The result of this situation, which has built up over that period, is a contradiction between what the Canadian Forces are *expected* to do and what they are *funded* to do and capable of performing. It is as if a school, for example, were allowed to deteriorate physically, its furnace worn out, its windows broken and not replaced, its teachers neglected, and the textbooks out of date. The parents and others would be up in arms and demanding an improvement. Yet Canadians and their governments during this period of relative peace have preferred to assume that all is well with defence and that its problems could be ignored.

The Canadian Institute of Strategic Studies, in its Strategic Forecast for 1990, observes that, in spite of the brave words of the White Paper "the record of events in 1989 depict a country lacking in the political leadership and national will to protect its own interests and sovereignty. This disregard was most clearly demonstrated by the April 1989 Federal Budget which effectively destroyed the commitment previously outlined by National

Defence Minister Perrin Beatty in 1987 'to help ensure for our children a sovereign and free Canada in a more peaceful world.' Within the space of a few pages of a budget report the government abandoned its highly publicized campaign to rebuild Canadian defence and foreign policy."[6]

Testimony to SCONDVA

The House of Commons Standing Committee on National Defence and Veterans Affairs (SCONDVA) heard many excellent presentations during 1989-1990. Among the witnesses appearing were General Gerard Theriault, former Chief of the Defence Staff, Mr Alex Morrison, Executive Director, Canadian Institute of Strategic Studies, and Mr Roger Hill, Director of Research for the Canadian Institute for International Peace and Security. General Theriault told us that

- there is a lack of a solid core to our own defence policy,
- we have excessive fragmentation of our forces (and) too many commitments,
- we have the longest coastline in the world, but only 18 maritime surveillance aircraft, which is not many more than the Dutch; and the Japanese have 100,
- our forces in Europe are no longer serving any vital military purpose, and
- Canada is already largely disarmed[7]

Mr Morrison contended that

- it is in the interests of Canada as a sovereign country, in the interests of the NATO alliance, and in the interests of international stability to show a determination to be a sovereign nation so that it can contribute effectively to its own security
- our maritime forces offer Canada an incomplete surveillance capability, incomplete capability to look after the sea lines of communication, and absolutely incomplete underwater capability, and
- that Canada risks being viewed increasingly as a protectorate of the United States.[8]

[6]"Introductory essay by Alex Morrison, p xxi
[7]SCONDVA Jan 25, 1990 pp. 6-8
[8]Ibid, pp. 10-12

Roger Hill observed that, whereas the Soviet Union had lost the capability of intervening effectively in Central Europe, some Canadian forces should be retained in Europe. Whether for collective defence purposes, or as guarantee or in a verification function, this "is one of the best ways of keeping an independent voice in alliance and European affairs and...helps to bolster this country's sovereignty. In the same way continuing to contribute to alliance naval forces on the northwest Atlantic and elsewhere can serve the same purpose..."

He added that "In today's rapidly changing world there is a serious possibility that NATO will be transformed in the next few years into a new arrangement having at its core a bilateral arrangement between the United States and some form of united Europe. Canada will have to play its cards skillfully if it does not want to be deprived of a major multilateral outlet for its diplomacy, marginalized from Europe, and thrust into an increasingly tight bilateral partnership with the United States."[9]

He also advised that

- arms control agreements are an effective way of promoting Canada's interests as well as security in the Arctic
- naval peacekeeping should be considered

Officials of the Department of Fisheries and Oceans also appeared before SCONDVA, and made valuable contributions to the committee's grasp of the maritime dimensions of Canada's defence problem. For example,

- coastal patrol activities around Canada cover *one million nautical square miles*
- the fishing effort there amounts to *180,000 days* per year
- the Atlantic fishery involves 12 nations within the 200 mile limit, 125 foreign vessels, and 80 separate foreign fishing quotas ; and in the Canadian zone the effort consists of 6,500 fishing days per year
- there are almost 800 vessels in the domestic fleet alone
- the Pacific coast foreign fishing is concentrated on the continental shelf, within Canadian fisheries waters, with two sensitive boundary areas at opposite ends of Vancouver Island.

[9]Ibid, p 13

To meet the challenge of patrolling and controlling our coasts and our fisheries resources, Canada has the facilities of Fisheries and Oceans, National Defence, the Coast Guard and the RCMP. The civilian fleet amounts to 323 vessels, mostly small, and can call upon the navy when required. The twin-engine Tracker aircraft of Maritime Command were used for inshore surveillance until being phased out, for reasons of economy, in early 1990. This service has now been replaced by private contract arrangements and by limited use of the Aurora long-range patrol aircraft. As there are only 18 Auroras in service, and three lightly-equipped Arcturus aircraft, due to arrive in two or three years, and as they are already stretched to the maximum of their potential usage, the prospect is not a cheerful one.

Brief descriptions of equipment holdings by the three branches of the CF are given in Chapter IV of this volume.

Defence policy - at least in theory - is drafted by defence specialists and bureaucrats, established by governments, paid for by the people. Where is Canada's defence policy today? Where is the "steady, predictable and honest funding program based on coherent and consistent political leadership" that was promised in the 1987 White Paper?

To say that Canada has problems would be an understatement. But many of our problems are self-inflicted, and the roots can often be found in our own weaknesses and fears. All too often Canadians and their governments live in fear of offending some group or another, and thus fail to take appropriate actions; in particular, we fail to declare to ourselves and to all the world exactly who we are, what we stand for and what our standards are. Out of fear of being thought racist or insensitive we have bent to accommodate to the wishes of newly-arrived immigrants - instead of introducing them to our Canadian ways and inviting them to live in harmony with us, while retaining their natural pride of origins. Out of fear of being thought warmongers we have bent to the sometimes uninformed pressures of disarmament groups instead of taking the actions we knew to be best.

Successive Canadian governments have made these mistakes. It is time we Canadians and our governments took a stand, declared our principles and invited others to join in upholding them. A good place to start would be in what ought to be the first responsibility of any government: the defence of Canada.

> *There can be no social security without national security*
>
> C.R. (Buzz) Nixon,
> former Deputy Minister, DND

> *Those who convert their swords into ploughshares may end up doing the ploughing for those who did not.*
>
> Anon.

II   DEFENDING CANADA

Does Canada need a defence capability? Why? and if so, what should this involve? And where should it be deployed? What about our membership in alliances?

A. Canada's Interests

Countries are not always governed consistently by principles, but they do have interests. Canada's interests, I would suggest, are oriented to peace and security with freedom and justice, a desire to enjoy the benefits of our geography and resources, and to practise cultural and economic relations with our trading partners and likeminded nations.

Recently we have come to recognize that our resources and those of others will and do require protection - both economic and environmental. Our concern for the environment must be translated into action. Good planets are hard to find, and we are trying to save this one.

Beliefs are also important; and to this Canadian there can be no belief more cherished than our belief in democracy, the essence of which is the freedom of choice. We are free to choose our political parties, our clothes, our spare time activities, our spouses. There is a choice, and it is ours.

We believe that democracy is not only a blessed condition for ourselves; we also believe that it has benefits for the rest of the world, even if indirectly experienced. Democracies tend not to go to war against each other. For a dictatorship, the decision to wage war is easily taken. The dictator speaks, and others obey. Who needs to waste time in debate? So it was, too, when the feudal system prevailed. The king would decide to make war on his enemy (or his prey) and would instruct his barons to provide the necessary forces. They in turn would be responsible for raising the armed peasantry and putting the whole assembled force at the disposal of the king. The result was sometimes a march to war for reasons we would have thought insufficient.

In a democracy, such decisions are not to be so lightly made, or by so small a group of decision-makers. And so the extension of democracy may well be an excellent way of encouraging the preservation of peace - especially when combined with sound policies for foreign affairs and defence.

Closely related to our belief in democracy is our faith in the open society, with our freedom to speak, to travel, to report the news, to enjoy all those other freedoms we have been able to take for granted - in short, to do whatever is not expressly forbidden.

Rounding out this catechism for a free country is our belief in human rights and the dignity of the person. We may not always have applied it perfectly in the past, but it is nonetheless a value for which we are prepared to struggle.

In coming to an understanding of security and what it means for Canadians, I would suggest that security lies in the ability to defend the integrity of our land and its approaches by sea and air, to uphold our values of peace and freedom, democracy, the open society, human rights and the dignity of the person.

In an essay on the great philosopher John Stuart Mill, Isaiah Berlin has written that "the defining proposition of liberal democracy is that it mandates means (elections, parliaments, markets) but not ends. Democracy leaves the goals of life entirely up to the individual." [1]

Then there is that troublesome word "threat" which has caused so much irritation to members of peace and disarmament groups in Canada during the past decade or two, although it is an extremely convenient word for military and strategic analysts of any persuasion. In a description of deterrence as a means of preventing war two Canadian analysts have observed as follows:

> " It is hard to overemphasize the psychological aspect of deterrence. J.D. Singer and the American behaviorist school define the perception of threat as the product of the estimated <u>capability</u> of the opponent's forces multiplied by the estimated <u>probability</u> that he will use them. Consequently there is no threat if there is no apparent intention to use force, even if there is clearly a powerful force available." [2]

In other words, threat perception involves *capability* as well as *intent*, so whereas Britain does not feel threatened by the US, both the US and the USSR have for many years felt threatened by each

---

[1] Isaiah Berlin in his essay John Stuart Mill and The Ends of Life, reproduced by Charles Krauthammer in the Washington Post, April, 1990.

[2] Albert Legault and George Lindsey in "The Dynamics of the Nuclear Balance", (Cornell University Press, 1974) quoting J.D. Singer, "Threat Perception and the Armament-Tension Dilemma", Journal of Conflict Resolution 2 (1958) pp. 90-105.

other. Happily, this perception is declining, although it would be foolish to leap to unjustified and premature conlusions about the continued need for prudent levels of defence. Canadians and other members of NATO are aware of the very high levels of military equipment being modernized by the Soviet Union. Even after the arms reductions that were agreed upon in Paris in November of 1990, that country will still possess enormous power to inflict damage on others if it should so choose.

B. Does Canada need defence?

Yes, it does, but not because of any warlike inclinations. The first responsibility of any government is to maintain a capability to defend the country and its people and to perform the functions of surveillance that preserve the sovereign integrity of its sea, land and air domains. To do otherwise might well put at risk the continued existence of that country. History is replete with suitable examples. As the cynic has said, "every country has an army in it; if not yours then someone else's". International pressures and restlessness are as inevitable as the bubbles in a boiling pot; and wherever there is dissatisfaction there will be the risk of wars and other forms of aggression.

Armed Forces have the potential to express the government's political will in the direction intended, and the possibilities are very wide: for example, coastal defence, fisheries patrols, aerospace attack warning, search and rescue, aid to the civil power, environmental surveillance, drug traffic interdiction or peacekeeping.

The uses of armed forces can be either offensive or defensive, and sometimes a mixture of the two. Faced with a military confrontation, the security problem of any country becomes quite obvious and urgent. Justifying the need for defence is then no problem at all. But more common is the use of armed forces for their symbolic presence - to prevent or to avoid the need for war. Aggressors from the time of Sun Tzu in China of the 5th century B.C. to Joseph Stalin in our own time have understood the value of a show of force, the means of winning without war.

But the peaceable kingdoms have also found the principle to be important for their more acceptable purposes. Since 1949 the countries of NATO, by the creation of an effective military alliance, were able to put an end to the territorial encroachment of a powerful aggressor. Stability was achieved. In fact, the period of the Cold War was a period of relative stability, at least for the members of the two alliances, NATO and the Warsaw Treaty Organization. They settled into the comfortable roles of familiar adversaries, each familiar with the capabilities of the other. This was a not intolerable situation for this country, although we would have preferred and have strongly advocated a significant reduction in the levels of arms while maintaining a controlled balance of power.

Now, in 1991, with the political situation changing in the countries of Central and Eastern Europe, and especially with the agonies of perestroika and glasnost in the USSR, new conditions are emerging. What will be required now are new approaches to the uses of defence forces for the preservation of international peace.

Further, Canadians must recognize that a military capability to fulfill any of its expected roles cannot be called into existence overnight. Development and construction of suitable new ships, aircraft or vehicles is a matter of years, and so is the training of the personnel - the men and women who would be expected to operate them. The recent example of the war in the Persian Gulf illustrates this point very well. Because Canadian personnel were well trained and accustomed to exercising with other NATO forces, they performed admirably. But because the ships were old and only barely adequately equipped, they had to be kept well away from the front lines. A fully equipped ship cannot be conjured out of thin air overnight. If Canada is to preserve its sovereignty and its security, a certain minimum capability must be kept in being at all times.

C. Is Defence Moral?

Whether defence is morally and ethically acceptable is or ought to be the concern of every thinking military person, as it has been with many thinkers of the religious and theological world. And when they have thought seriously about the question they are likely to conclude that it is indeed moral. When we consider the alternative, that of leaving a country undefended, the answer becomes obvious. Refusal to defend ourselves and our fellow citizens from the violence of a determined adversary would be reprehensible, an act of criminal negligence.

There are those who habitually advocate the dissolution or further reduction of Canada's armed forces. They appear to be motivated primarily by the selfish desire to divert funds to their own private causes, and ignorant of the potentially disastrous consequences of their policies. Among these consequences would be a reduced ability to assert sovereignty over Canadian domains, and increased reliance on the Americans for Canadian security. To an extent, this process has already begun.

On balance, therefore, the maintenance of national defence and the strategy of deterrence of aggression is morally acceptable, if only because the consequences of the alternative would be evil and disastrous for Canada, and possibly for others as well.

The trick is to steer a course between a prudent level of defence and one that would be provocative. By way of example, if Canada had a dozen submarines, capable of patrolling all our oceans, and 100 interceptor aircraft, this capability would be clearly defensive in nature. But if this country were to launch 400 submarines, and arm them with long range missiles, and build several hundred long range bomber aircraft, then our capability could be seen as provocative and therefore destabilizing. We have never yet been in that situation, and I doubt if any Canadian would wish it.

****
*War is an ugly thing, but not
the ugliest of things; the
decayed and degraded state of
moral and patriotic feeling
which thinks that nothing is
worth war is much worse.
A man who has nothing for which
he is willing to fight, nothing
he cares about more than his
own personal safety, is a
miserable creature who has no
chance of being free, unless
made and kept so by the
exertions of better men than himself.*
****

(Seen on a plaque in the
Officers' Mess at CFB Baden)

D. The Changing International Scene

Speaking of Changes in Eastern Europe, it is interesting to consider the following selection of headlines and quotations:

> *"What do we want of the Soviets?...We could want them to stop expansionism, to stop destabilization (of other countries), to stop stealing Western technology, to stop spying, to stop hijacking international organizations like UNESCO...to stop ideological warfare, to stop terrorism."*
> (Jean-Francois Revel, 1986)

and then:

> *"The idea that people should fight to death for one or another ideology should be once and forever forgotten. The values of peace, prosperity and humanism are more important than any ideology."*
> (Vadim Medvedev, Ideology Chairman, USSR reported in Pravda, October 5, 1988)

> *"Shevardnadze denounces Afghan war, admits ABM treaty violation"*
> (Globe and Mail October 24,1990)

> *"Belgium, Hungary sign 3-year pact" (for limited military co- operation)*
> (Globe and Mail November 6,1989)

> *"Gorbachev calls for return to spiritual values"*
> (Globe and Mail December 1, 1989)

> *"The despotic authority is in the final count a very insecure power"*
> (Novosti February 26 1990)

"We should abandon the ideological dogmatism ... and outmoded views on the world revolutionary process...(and) abandon everything that led to the isolation of socialist countries (and) the understanding of progress as a permanent confrontation with a socially different world."
>(Gorbachev's address to
>Communist Party Central Committee,
>February 1990.(G&M)

"Poland, still a member of the Soviet-led Warsaw Pact, has spoken out in support of German membership in NATO (as) neutrality would isolate Germany"
>(Globe and Mail editorial,
>March 20, 1990 reporting speech
>of K. Skubiszewski,
>Foreign Minister of Poland)

"Soviet Soldiers of Occupation Seek Asylum in Czechoslovakia"
>(The Independent, Feb.21,1991)

"Havel Seeks Closer Ties Between East and NATO"
>(Washington Post, March 22, 1991)

"Warsaw Pact Nations Sign Formal End to Military Alliance"
>(Ottawa Citizen, Feb 25, 1991)

HOWEVER, let us also consider these comments:

*"Is Moscow Cutting Its Military? No, It's Building Up"
...The Kremlin recently began to deploy a new ICBM"
"Soviet production ... rate for main battle tanks ... was 3,400, up from 2,800 a year from 1982 to 1984"*
>(Frank J. Gaffney Jr.
>NY Times Nov 17 1989)

"Launch of new Typhoon sub" (headline in Globe and Mail, Jan 30,1990, reporting that 6 of the giant ballistic missile-carrying are now based in Kola peninsula)

"Group Cites Continuing Abuses in USSR"
(Ottawa Citizen,June 5, 1990)

"KGB Still Monitoring Non-conformists, ex-officer charges"
(Ottawa Citizen,June 17, 1990)

"Troops Raid Nine Lithuanian Buildings"
(Ottawa Citizen, April 26,1991)

AND we might also consider a few other sobering facts: while the democracy movement has been surging through Central and Eastern Europe, it presents a challenge not only to the Soviet empire but also to Canada and the other countries of the West. Easterners are unaccustomed to democracy with its rights, its obligations and its costs. In some countries the people seem to be placing too many of their hopes on a single person - merely transferring their allegiance from a sullen acceptance of the former dictator to an unreal expectation of the new and more popular leader. It is unlikely that, for example, Vaclav Havel in Czechoslovakia or Lech Walesa in Poland could ever meet those expectations without the people themselves becoming more active participants in the process.

On the economic side, the adjustment is proving just as difficult for those countries. For more than 40 years the countries of the socialist bloc lived under a system of fixed prices for consumer goods. Everyone knew the price of a kilo of potatoes or a bottle of milk. To the average citizen it was one of life's comfortable certainties. But now, in countries where most of the people have never had bank accounts, do not know the difference between price and value, the uses of credit and the benefits as well as the risks of competition, there is bound to be unhappiness. And where there is unhappiness, instability may well follow.

And so we are not at the end of a turbulent time. Far from it. We are probably on the threshold of a long period of political unrest. Strong hands will be needed at the helm. Strong organizations, strong institutions such as the North Atlantic Treaty Organization (NATO) will find new meaning and new responsibilities in providing the guidance and stability that these emerging democracies will need so badly. The challenge of maintaining stability in Europe will be greater than at any time in the past 20 years.

We are entitled to be optimistic; we are obliged to be cautious.

E. Do We Need Alliances?

1. Lessons Learned

OK, so defence is important. But why must we belong to alliances like NATO and NORAD? Do alliances cause wars with their entangling commitments? Is there a chance that we could set a good example by being strictly neutral?

The answer is that it would cost us more to be neutral, we would be less secure, and there is no reason to believe that such isolation from the ebb and flow of international affairs would have any beneficial impact whatsoever on the risks of war and other forms of aggression. Take Switzerland, for example. Today that country is better known for the selling of arms than for negotiating peace.

Neutrality means being alone. Neutrality also carries obligations in international law, such as the obligation to defend one's territory and to prevent it being used by a belligerent power. The neutral states of Europe may appear to have it both ways: freedom and peace, provided by the protection of the western security system, and with no entanglements. But not without cost. These costs can be measured in various ways, but it can be seen by the interested observer that neutrals like Sweden and Switzerland are paying for their presumed independence and freedom of action.

Table 2

SOME COMPARATIVE COSTS of NEUTRALS vs ALLIES

	% of pop'n in Forces	Numbers in Forces incl Res. (in 000's)	% of GDP for defence	% of gov't budget	$US per cap.
SWEDEN	9.3	676	3.0	8.8	364
SWITZERLAND	9.8	605	1.9	19.7	282
NORWAY	7.7	236	3.3	6.7	426
USSR	4.0	11,313	14-17*	?	?
USA	1.4	3,839	6.4	27.1	1,061
BRITAIN	1.1	636	4.7	10.9	402
CANADA	0.4	137	2.1	9.8	307

It is interesting that Switzerland, with a population of only 6.5 million, has almost as many troops on call as Britain. Most are well-trained Reserves.
Source: International Institute for Strategic Studies, The Military Balance 1987-1988 and 1989-1990.
* Mr. Schevardnadze estimated, in 1989, that it might be as high as 19% and, more recently, that it might be one quarter of the total (Ottawa Citizen, July 4, 1990 p.1)

The neutrality option does not exist for Canada. Although there is no direct military threat to this country, we face dangers nonetheless. Geography has placed us between two great powers, both of them armed to the teeth; and it is very much in our interest to ensure that they never come to blows. Is it any wonder, then, that we feel a global responsibility? Large wars, or even medium sized conflicts could endanger us as well as others. We cannot be conquered, but we could be destroyed.

Neutrality has never been a viable option for Canada, from our earliest colonial times to the present day. With the second largest land mass, and with the longest coastline in the world, we have a country that is difficult to defend. With our large size and our small population it can be difficult for us to know when encroachments occur. Then, even when we have that information we have the further problem of providing the military forces with which to repel or escort intruders from our shores. If we intend to maintain our sovereignty claims over the entire surface of Canada then we must possess some semblance of capability to enforce these claims.

With the help of our allies we can hope to maintain a credible defence at less cost than if we were to try to go it alone. But it is not only for selfish reasons that we seek our security in alliances; it is also more effective for all concerned.

Collective security, properly maintained, can ensure that weak countries are not picked off one by one, as Hitler was able to do with the frightened countries of Europe in 1939 and 1940. Because they were sick of war, these countries tried to ignore the bellicose threats of the Nazis. It was the wrong policy, and they were all the losers as a result. It is tempting to wonder what the result might have been if there had been a NATO (or, for that matter, a Warsaw Pact - as the Warsaw Treaty Organization is usually called) in 1939.

Canadians should take a hard look at the lesson of the peace movements of that time. They were opposed to both fascism and war, but they believed that by declaring their opposition to both that they had solved the problem. They had done no such thing, as the events of 1939-1945 were to prove so tragically.

2. Accepting Responsibility

There is an added reason for Canada to subscribe to collective security, and that is our feeling of global responsibility. Instead of remaining selfishly aloof from the problems of other countries, we Canadians have cheerfully assumed our global responsibilities. We demonstrate this continually in our aid to Third World countries, in our peacekeeping operations wherever they are needed, and in our responsible, prudent efforts to maintain peace through collective security.

War is all too common. In their classic study "The Lessons of History" the historians Will and Ariel Durant reported that, of 3,421 years of recorded history, they could find only 268 years that had seen no war. We probably cannot hope realistically to prevent war everywhere on earth, but we can hope to contain it, to prevent it from spreading and escalating. Not for us the luxury of staying insulated from reality behind a wall of indifference to the sufferings of others.

We are a trading nation, and a caring nation. In the past we have joined in causes that were just, in the hope of restoring peace and freedom. Today we are no less concerned, and we show it by participating in peacekeeping operations in less fortunate lands, by cooperating with our allies in maintaining the freedom of the seas, and in other related ways.

We are active in promoting the success of arms control and disarmament. We perform these functions through a variety of organizations, principally: NATO, NORAD, the United Nations, the Conference on Security and Co-operation in Europe (CSCE), the Force Stability Talks, and the recently concluded Conventional Forces in Europe Talks (CFE) at Vienna.

Membership has its privileges, goes the saying. And because we are members of these organizations, some of them having a military capability, we have the right to a seat at the negotiating table, a right to be heard and an effective means of using that right. This allows Canada to influence the ways in which nations and organizations address political and economic problems. It forces others to look at Canada's concerns; and it allows us to influence our allies . The restraining effect on an ally should not be underestimated.

3. NATO - An Effective Alliance

Speaking of alliances and military capability, some Canadians have speculated about dismantling both NATO and the Warsaw Pact. In any such discussion it is important to recognize that the two are not alike. That they both have possessed a military capability is obvious, but there the comparison ends. In NATO the military and political decision-making is democratic, the various committees meet frequently, and the strongest power does not always get its way. This was never true of the Warsaw Pact, which until its dying days was very forcefully dominated by the Soviet Union. As of March 31,1991 the military structure of the Warsaw Pact has passed out of existence, leaving little else.

Beside the military aspect of NATO, which is well known, there is a political and economic dimension that will grow in importance during the next few years. The closest counterpart in the Warsaw Pact would be the Comecon, a rigid system of forced trading and artificial pricing which is now disintegrating under the pressures of Eastern European perestroika and the imperatives of the market economy.

Then there is the third dimension of NATO, the scientific affairs cooperation and scholarly exchanges, and the cooperation and coordination of international projects to find solutions to environmental problems. Article 2 of the North Atlantic Treaty, which was included at Canadian request, provides that

> *"The parties will contribute toward the further development of peaceful and friendly international relations by strengthening their free institutions, by bringing about a better understanding of the principles upon which these institutions are founded, and by promoting conditions of stability and well-being. They will seek to eliminate conflict in their international economic policies and will encourage economic collaboration between any or all of them."*

In an age when we are discovering the enormity of the crimes committed against the environment of land, air and water, especially in Eastern Europe and in the oil fields of Kuwait, NATO is quietly cooperating in order to find solutions to these problems and make them available to the world. And NATO countries are assisting each other to improve their economies. Between the two alliances there have never been many points of meaningful comparison. What the future holds for the remaining shreds of the Warsaw Pact may be very little; it never had much to offer to the world except weapons and threats, and in recent years has become very unpopular with its own members, and so is likely to disappear utterly except for the unpleasant legacies of pollution and the bitter memories of political oppression.

NATO, on the other hand, has proven to be a success and is poised on the verge of extending the benefits of that success to others. Forty years of peace has been no mere coincidence.

The Declaration of the Heads of State and Government, at the meeting of the North Atlantic Council in London on July 5 and 6, 1990 justifiably proclaimed NATO " the most successful defensive alliance in history."

As mentioned previously, NATO is about to re-align its roles, putting less stress on its military functions and more on the political, economic and "third dimension" activities. Although many people in Canada and elsewhere may have been unaware of the fact, NATO is more than just guns that point East. NATO has for many years had important political, economic and scientific dimensions - with quiet achievements that may have been some of the world's best kept secrets.

How many of us know of the study on Advanced Waste Water Treatment, conducted jointly by the United States, Canada, Britain and Germany ? The study resulted in assistance to member countries to develop more effective and efficient means for purifying water. And how many know of the Canadian-led study on Nutrition and Health, or the German-led study into the preservation of stained- glass windows that led to development of a protective coating to protect historic windows from the ravages of acid rain and other pollutants?

Now that the military requirements are being reduced these and other non-military roles will come to the fore; the existing strong mechanisms developed over the years will enable NATO to extend the benefits of its knowledge and its collective resources to other countries, including those of the Warsaw Pact. A family of nations whose values have triumphed, whose energies and ethics have brought peace and prosperity to its citizens, NATO is now poised to help others achieve the same benefits - if they want them - while also preserving the political stability of Europe.

It is persuasive testimony to the moral power and acceptability of NATO that countries of the former Warsaw Pact should wish to join it. In extending a polite refusal to Czechoslovakia and Hungary in April, 1991, Secretary-General Manfred Woerner said that NATO is not seeking to extend military borders to the east; but he added that "What NATO can offer is a multiple and extensive web of relations which does not exclude the Soviet Union but wants it to be a constructive and creative partner as well."

Canada must continue to be a proud member of NATO. The expression that membership may take, especially in its military dimension, is another matter and one that offers much opportunity for constructive debate.

The easing of East-West tensions - assuming they continue to reduce the need for weapons - may not affect Canada directly. In the total comparison of forces available or in Europe, ours are miniscule; and, in any event, we have no nuclear weapons at all, at home or abroad.

Our presence in Europe is partly symbolic for the unity and resolve of the NATO alliance. The advantage of being there, however, is that we are able to exercise together, to work together and to communicate together among sixteen nations. And this includes being involved and consulted in discussions for the reduction and control of arms. This is not an inconsequential benefit in an increasingly interdependent world.

As in so many matters, it may be that others are able to recognize the value of our contribution to NATO more readily than we do. A distinguished German authority has said of Canada's role:

> *"There can be no doubt that the Europeans wish to see an undiminished Canadian participation in the North Atlantic Treaty and in European affairs, not only for our own sakes but also for the sake of Canada herself."* [3]

As for the level of our contribution to defence through NATO, Canada is getting a cheap ride. In the mid-1960's defence accounted for some 24% of the federal budget. In 1990 it is closer to 10%. Successive governments have reduced spending on defence, as a proportion of total spending, until only a slender minimum of credible defence has been left; and Canada's Armed Forces today are unable to afford necessary replacements for worn-out equipment.

WE FIGHT NOT FOR GLORY
NOR FOR WEALTH NOR HONOUR
BUT ONLY AND ALONE WE FIGHT FOR FREEDOM
WHICH NO GOOD MAN SURRENDERS
BUT WITH HIS LIFE.
The Declaration of Arbroath
1320 A.D.

---

[3] Heinz von zur Gathen, retired Generalleutnant of the German Federal Armed Forces, former NATO official and regional chairman of the Society of Military Science, "Canada's Role in Europe", MacKenzie Paper No. 10, 1988.

III CONSIDERING THE OPTIONS

A. Our Present Situation

Canadian Defence policy flows from Canadian security policy, a seemingly tidy two-step approach, assuming that security policy is part of a coherent foreign policy. Here is the security policy, as framed by the Department of External Affairs:

1. Prevention of war and the deterrence of aggression through collective defence arrangements of NATO and NORAD
2. Pursuit of effective and verifiable arms control and disarmament agreements
3. Commitment to the peaceful settlement of disputes and collective efforts to resolve the underlying causes of international tension

For more than 35 years these three pillars have been the foundation of Canada's security policy. They still make good sense; and members of the Department of National Defence who have taken community speaking engagements in recent years have reported widespread public acceptance of the policy by the men and women of Canada they have met in their communities. Canadian public acceptance was even more evident when, on the occasion of the brief war in the Persian Gulf, the policy was translated into action.

Defence policy rests on five pillars, all flowing from the security policy and expanding upon it:

1. Strategic deterrence
2. Conventional (non-nuclear) defence
3. Protection of national sovereignty
4. Arms control and disarmament
5. Peacekeeping

It would be fair to say that, while our policy is built in part on collective security, an important part of our strategy is deterrence. Our greatest security fear has been of global war, especially nuclear war. From such a disaster this country could not hope to escape. It would matter little how pure our actions, how many years had elapsed since we last possessed a nuclear weapon, or how friendly we might have been to other countries. None of that could save us if ever such a conflagration should erupt. And so we Canadians must spare no effort to ensure that this will never happen.

That is a good argument for self-interest, but Canada is not motivated by pure self-interest alone. We Canadians have a long history of caring, of a sense of responsibility for the fate of others. It has been manifested in many ways. In earlier times, we went to the aid of the mother country in her time of need; we sent our missionaries abroad; we opened our doors to refugees. Today we also exercise a sense of responsibility. Through our security and defence policies we strive for the preservation and enhancement of international peace for all peoples, which is in many cases the precondition for freedom and prosperity.

It has been said that Canada's eastern border is not on the Grand Banks; it is in North Norway and in Central Europe. This approach carries two implications. One is that we are serving to keep whatever fighting there may be far away from our own borders. More importantly, we are also ensuring that wherever stresses may build up they can be quickly smoothed and subdued before they erupt into full-scale wars. We have experienced two world wars in this century, and we are not alone in thinking that this must not be allowed to happen again. As responsible international citizens we practise prevention; and an important element of prevention is deterrence.

A Word About Deterrence

Deter: to discourage or hinder (from) by fear, dislike of trouble, etc. This is the definition given in the Concise Oxford Dictionary. Deterrence is a strategy that NATO countries have practised for many years, and it works. There are many analogies, and each of us has our favourites. For example,

- we lock our doors at night or when we go away, to deter crime,
- we set alarms, to discourage intruders,
- the cop on the beat, by his very presence, deters crime,
- the bully in the school yard won't pick on the kid who seems likely to fight back,
- football players rarely get mugged.

Deterrence works. But in confronting the determined doubter, it may seem almost impossible to prove that it works. The doubter may suggest that the non-aggression of the Soviet Empire against the Federal Republic of Germany since 1949 could have some explanation other than the determination provided by the united countries of NATO to preserve their freedom. Some may also believe that the USSR might have voluntarily given up their SS-20 LRINF missiles without the deterrent pressure of the NATO Cruise and Pershing II counter weight. Perhaps these achievements could have come about through some spontaneous miracle without the efforts of NATO. Perhaps. But surely this is grasping at straws, a search for a way to avoid having to admit the value of deterrence in preserving the peace.

In his book "Peace With Freedom" Dr. Maurice Tugwell, Director of the Mackenzie Institute, puts it this way:

> *"Deterrence means ...the deterrer seeks to ensure that whatever military action or political bullying a potential aggressor might contemplate, he could not foresee any likely situation in which victory is obtainable without costs so unacceptable that the word victory loses its meaning. Nuclear weapons combined with modern means of delivery have transformed the notion of deterrence from a pious hope to a reality."*

The strategy of deterrence is a common sense approach to keeping the peace. It is one element in a posture of preparedness which is usually phrased as one half of a double-barrelled policy: to prevent war by a strategy of deterrence but, if deterrence should fail, to be able to defend our country and its interests. But a government must ever be mindful of the risk of overdoing it.

Deterrence must never be allowed to become so great that it becomes a provocation to others. Increasing one's own security to the point that the neighbours become insecure is no way to preserve peace. That is the mistake the USSR made in maintaining massive military superiority in Europe since 1945, which caused others to become alarmed; and that is a pitfall the USA must bear in mind in its current and future negotiations for arms reductions in Europe and elsewhere, now that the balancing strength of the Soviet Union has been weakened.

A few years ago, when the Cold War was very much on, a joke was making the rounds of the diplomatic community: "to a Russian, the only safe border is one that has Russians on both sides of it". This neatly summed up the paranoid fear for their security that produced feelings of insecurity in others.

The trick is to have just the right amount of deterrent, a prudent level of ability to defend ourselves and, as Lord Carrington used to remind us, the demonstrated willingness to use it if that becomes necessary.[1] Of course, an excess of deterrence has never been a problem for Canada and it seems unlikely to become a problem in the foreseeable future.

Preparedness

Does Canada really need to be prepared to defend herself? I suppose it all depends on our attitude toward defence, and toward preparedness in other ways. Do we make provision for a financial future? Or do we just hope that something will turn up? Do we plan to have the family car serviced occasionally? Or do we just run it until it collapses? Are we to be like the carefree grasshopper of the ancient fable, or the more prudent ant who laid up stores for the winter? These needs and our attitudes toward them will be determined by how much of a gambler we think we are. No war on the horizon? Then why bother with defence? For some persons it may seem as simple as that; but not, I suggest, for those who are serious about security and peace.

In matters of security one of the important lessons of history is that we cannot easily predict where the next challenge is coming from. In the late 18th and early 19th centuries the Barbary pirates were the scourge of the seas and a plague on "such as pass on the

[1] Lord Peter Carrington, Secretary General of NATO, 1984-88. NATO Review, June 1984.

seas upon their lawful occasions".[2] In recent times we have seen the bombing of an Air India flight by Sikh extremists, the bombing of a Panamerican Airways flight over Scotland, presumably by the PLO, the destruction of relief food supplies in Africa by warring factions, and so on. It is of such incidents that larger wars are sometimes made. And so we Canadians have long ago determined that it is sensible (and cheaper in the long run) to always maintain a modicum of preparedness for security emergencies. The key question is not "whether to" but "how much". We must expect the unexpected.

Continentalists or Internationalists

I have said that Canadians consider themselves to be globally responsible citizens. But there is a choice. We could get completely out of Europe and concentrate our defence capability in North America. Our CF-18s could certainly be put to good use in northern patrols; our diminutive fleet of ships would have their hands full patrolling our coasts, even without ever venturing out to the high seas; and our ground forces have always conceded that even Canada alone is too much for their modest numbers to protect unaided. It has long been a joke that Germany keeps more tanks for training purposes in Canada than Canada's entire complement of Leopard I tanks.

Of course, we could always count on the Americans to look after us, couldn't we? Already they have the ability to cruise our arctic waters undetected by us, to play cat and mouse with the Soviet navy if they wish. They can fly across our borders, and they could even send in troops if we need them (or if they think we need them: there might be a distinction). In fact, there are contingency plans for the movement of U.S. troops here, to defend vital installations such as the Saint Lawrence Seaway if needed. This is part of the plans worked out under the Canada-US Military Co-operation Committee.

How would Canadians feel about having Uncle Sam as our only ally? NORAD our only treaty of mutual defence? Canadians would not care for such a smothering relationship, and it would be a quick route to a loss of sovereignty. No thanks! We like the Americans, but we really feel more comfortable having other

[2]from the Naval Prayer.

friends as well. That being the case, we have a choice of international groupings to which we might belong. There is the United Nations, the European Community(EC), the Conference on Security and Co- operation in Europe(CSCE), and there is NATO.

First, the EC: it is a European organization, drawn together from the European Coal and Steel Community, the European Economic Community and the European Atomic Energy Community. It became a united entity on 1 July 1987. Concerned with its own problems of political unity - including the 1992 target for completion of the EC's internal market - it has not yet worked out a satisfactory defence-industrial policy for Europe. Its focus is Europe; its members are Europeans; it does not include Canada nor is it ever likely to want to do so. The EC is not for us.

Second, there is the CSCE: like the EC it is Euro-oriented, the difference being that it includes all NATO and former Warsaw Pact countries and therefore Canada is already included. Under the Helsinki Final Act the CSCE addresses three "baskets", the first of which holds questions relating to security in Europe. The other baskets are co-operation in Economics, Science and the Environment, and co-operation in Humanitarian and other fields. In terms of its first "basket', CSCE is a forum for arms control negotiations; and it has produced some very encouraging progress since the 1986 Stockholm agreement for confidence and security building measures. But CSCE is not an organization that pretends to coordinate, let alone create, military forces.

In a speech given in Toronto in May of 1990 Joe Clark observed that the CSCE has functioned until now on an intermittent basis, and he expressed the hope that it might in future be able to institute on-going co-operation in security, in verification and confidence-building, in human rights, economics and the environment. He also suggested that "NATO and the CSCE are the two complementary institutions in building the new Europe". [3]

Third, there is the UN: A very large organization, having more than 150 members and forced to make many compromises, the UN has not been capable of coordinating military forces to meet an emergency. The action that sent forces into Korea in 1950 with blue berets is not likely to be repeated in our lifetimes. A glance back at the financial and moral problems that have plagued UNESCO, the UN specialized agency for Education and Scientific Cooperation, gives an inkling of the UN's incapability and limita-

---

[3]The Hon. Joe Clark, Secretary of State for External Affairs, speaking at Humber College, Toronto, May 26, 1990 on "Canada and the New Europe".

tions. The capacity to lead a coordinated force for a precise purpose, such as the stopping of aggression or the disciplining of a member country is just not there. Peacekeeping operations are another matter, and Canada is a regular and very valuable participant in those.

Yet, having said that, I must admit to being pleasantly surprised that the UN was able to assume leadership during the Gulf crisis of 1990-91, enough to allow it to authorize military action by member states in order to repel aggression and to restore the independence of Kuwait. In fairness, though, it must be added that the reason Coalition forces were able to function so smoothly in the Gulf theatre was that so many of them were NATO members, accustomed to working together and rehearsing together, with standard procedures worked out over many years of co-operation within the Alliance. NATO provided the well-oiled mechanisms for the supply of fuel, food, communications,the servicing of ships, aircraft and vehicles - and all the other functions that are necessary to a well-run operation. The NATO team was the core of this very successful operation, the glue that held the Coalition together and enabled it to do the job as directed by the UN.

Most years, members of Canadian Forces take part in a variety of exercises with NATO and NORAD allies. This is the way a good defence force keeps in trim, the way it maintains and improves its readiness, should its country need it in time of emergency. For example, the navy participated in maritime coordinated exercises (called Marcots) with British, Dutch, American and French ships, submarines and aircraft in 1990. Ground forces based in Europe took part in a Divisional exercise called Royal Sword, with our German allies; and Canada's air forces are hosts every year to US air forces in Maple Flag, a NATO air combat exercise. Details of these and other NATO and NORAD exercises in which Canadians participate regularly may be found in the DND publication "Defence '90"

That brings us to NATO itself, the North Atlantic Treaty Organization. The appeal of NATO is that it already exists, it has been successful, and it carries many other benefits than those of a purely military nature, although these values were proven in the recent Gulf conflict. Its 16 countries have many features and values in common. With varying degress of success, they are all dedicated to democracy. And it is precisely for these reasons that I have faith that NATO also has the innate flexibility to adapt to changing circumstances in the 1990s and to contribute to the stability of Europe, which is so necessary to the cause of peace and security.

We know that the Europeans want us to remain firmly in NATO, and so do the Americans. But we are not in NATO purely to please others: we are in NATO because it suits our interests to be in NATO.

Policy Making

Defence policy is made in many places: in the department of National Defence, in External Affairs and in Cabinet. The basic security policy, for which External is the lead department, is sound and continues to be acceptable after many years of consistent application. DND has also produced strong policy analysis and advice to Ministers. Unfortunately, it is the political arm of government that has occasionally seemed guilty of a certain aimlessness, leaving Canadians with the uneasy feeling that haphazardness and ad hocery have been holding sway. Someone needs to tell someone, I believe, to stop all this swaying and develop a steady course for the future.

The Standing Committee on National Defence and Veterans Affairs (SCONDVA) of the House of Commons provides an excellent forum for the questioning of expert witnesses, and as a member of SCONDVA I have been extremely impressed with the variety and quality of informed advice that is available to us. Much of that quality advice has found its way into this volume; and to those who appeared before us in our hearings and who provided such excellent responses to my questions and those of my colleagues I am deeply indebted.

Special committees are also formed from time to time, to consider special defence problems. For instance, the problems brought about by the introduction of cruise missile testing in Canadian airspace in 1983-84 resulted in the overnight development of an interdepartmental committee of analysts and information specialists from National Defence, External Affairs, the Privy Council Office and the Prime Minister's office.

Still, the feeling persists that Cabinets are vulnerable to the transient pressures of lobby groups, and that serious decisions on matters such as high technology submarines or low level air defence systems may be swayed by the perceived concerns of noisy pressure groups whose grasp of the subject may be shaky at best.

What is needed to counter this impression and to ensure that governments have access to advice that has been analysed, digested or presented raw - as the client wishes - is a Council whose sole mandate is to advise the government on sovereignty and security policy. Such councils already exist in other spheres, such as the Economic Council of Canada. But no one model need be followed. In fact, the Minister of National Defence already has an Advisory Council on Defence, recently appointed, and this author is its Chairman. The present Council, however, is restricted to members of the Progressive Conservative Party caucus.

> Serious thought should be given to creating either a pair of broad, non-partisan Councils to analyse and digest advice from many quarters, and to advise Cabinet as required – one for the review of foreign and security policy and one for defence policy – or to a single Security Council with separate sub-committees for foreign and defence policy. Members should represent a variety of related expertise and a mix of the political and non-political.

B. The Pillars of Defence Policy

1. Nuclear Deterrence

Nuclear deterrence is the third leg in the triad of NATO deterrent strategy:

a. Conventional weapons
b. Theatre nuclear weapons
c. Strategic nuclear weapons.

In practice, this means that, if NATO countries were under attack NATO would attempt first to respond by resisting with conventional (i.e.non-nuclear) arms. If that response failed, NATO would then attempt to stop the aggression by means of short-range or "theatre" nuclear weapons. Third, and only in the event of a continuing and serious aggression would resort be made to the long- range, or strategic, nuclear weapons.

In light of the encouraging developments of recent months, NATO is already re-thinking this strategy with a view to de-emphasizing the nuclear components. With relations between NATO countries and those of Eastern Europe improving as they seem, the prospect of NATO ever using theatre nuclear weapons becomes less likely. The unifying forces in the two Germanys during the late 1980s brought this into focus, even before Foreign Minister Hans Dietrich Genscher and others had observed that the shorter the range of the weapon, the deader the German.

In May of 1990, the U.S. Defence Secretary announced that there would be no modernization of the short range Lance missile. As a result of these encouraging developments, we can now foresee a de-emphasis on the second leg of the triad, the theatre nuclear deterrent, thus offering an illustration of the nature of effective deterrence and its needed flexibility to be scaled up or down in response to the perceived challenge. At their summit meeting of July, 1990 NATO ministers agreed that, from now on, nuclear weapons would be retained for use only as a last resort.

Canada does not possess nuclear weapons nor is there any intention of our acquiring any. At one time this country had these weapons, but the last of them disappeared from our soil after the CF-18 fighter aircraft was introduced in 1982, replacing the nuclear-armed CF-101 Voodoo.

Canada's NATO and NORAD allies have the nuclear deterrent, of course, and Canada benefits from that deterrent protection. Canada also contributes to that deterrence at the strategic level by our participation in joint programs of surveillance, warning and training of strategic forces, nuclear or otherwise. As Mr. Beatty stated in the 1987 White Paper:

> *"we enhance deterrence to the extent that we are able to deny any potential aggressor the use of Canadian airspace, territory or territorial waters for an attack on NATO's strategic nuclear forces."*

He also said:

> *"Stable deterrence at the strategic level is essential to the security of Canada. The Government will continue to contribute to the maintenance of an effective Allied deterrent according to our own independent analysis of the strategic environment."*

Other ways in which we contribute to strategic deterrence, some of them not nuclear weapons-dependent, are by participating in the NATO Airborne Warning and Control System (AWACS) and by making our training and testing facilities available to our allies. In the words of the White Paper,

> " Canadian airspace and military ranges and training areas are also used to test and evaluate the performance of Allied weapons, most notably the United States air-launched cruise missile. Some of these are nuclear capable, but no nuclear weapons are tested in Canada. Warships of our allies regularly visit Canadian ports. Such visits are frequently made on the occasion of exercises during which Allied ships , including Canadian warships, practise combined operations. They are a logical consequence of our membership in an alliance and of our acceptance of the protection offered by collective defence."

However, most of us would feel more comfortable with a lower level of such weapons. It is difficult to escape the knowledge that, wherever we travel, we may be in an area that is targeted by these terrible devices. Cities I have visited in the countries of the Warsaw Treaty Organization are targeted by the weapons of our allies, and we in turn are targeted in many areas of our NATO countries by weapons belonging to the Soviet Union.

So, although supporting the value of the strategic deterrent I would like to see fewer of these weapons. But how many should there be? In a recent count it was estimated by the International Institute for Strategic Studies that the total number of strategic nuclear warheads is about 27,000 divided between the two superpowers as described in Table 3 below.

Table 3 Soviet-American Nuclear Balance

	United States	**Soviet Union**
ICBM	2,450	6,657
SLBM	6,208	3,806
Bombers	5,872	1,940
Totals	14,530	12,403 (warheads)

Source: IISS The Military Balance,1989-90, p.212

The merits and the terrors of nuclear weapons have been debated and pronounced upon by many expert authorities and by those who are not so expert. Some would like them to be eliminated altogether. But others feel safer because of their existence. And, besides, we know that nuclear weapon technology cannot be disinvented. The atomic device that exploded over Hiroshima created an emotional shock wave that travelled around the world. In a way, it is still reverberating. That blow was so shocking to knowledgable people, in many countries, that it may well have brought about an important measure of restraint. Armies are less likely to gallop into "glorious" battle if, waiting for them, are weapons of devastating power.

The 1945 bombing attacks on Hiroshima and Nagasaki are frequently cited as frightening examples of man's power of destruction - and rightly so. But we should also remind ourselves that these early bombs did not start a war: they ended one. Winston Churchill, in his memoirs, reminds us of the fierce and bloody fighting on Okinawa Island and elsewhere: " I had in my mind the spectacle of Okinawa (where) to quell the Japanese resistance (in Japan) man by man and...yard by yard might well require the loss of a million American lives and half that number of British - or more".[4] Taking into account the fact that the Japanese had lost 135,308 soldiers and 75,000 civilians in the battle for Okinawa alone, there would have been an enormous loss of Japanese lives before the Pacific war could be won by the Allies. American sources had estimated that, without the nuclear weapon, the lives of one million Japanese troops and two million Japanese civilians could have been lost before the Pacific war could be won by the Allies.

A question sometimes raised is whether the elimination of nuclear weapons (assuming it a possibility) might merely make the world safer for the practice of so-called "conventional" warfare. Since the Soviet Union has long had the preponderance of conventional strength, at least on land, this view has been held rather more widely in the countries of NATO.

The picture can be further confused by attempts to compare nuclear capability in terms of warheads (the US has more) or in terms of total megatonnage (the Soviets have more) or in a dozen other comparisons. The important matter, however, is that the number is still very high, even if declining slightly since the

[4]Winston S. Churchill, "Triumph and Tragedy" p. 638.

1960s. Canada is right to press for reductions on both sides.

A matter of even greater concern, however, is the very real fear of horizontal proliferation of these weapons. Officially only five countries are nuclear powers: USSR, USA, Britain, France and China. However, there are very real fears that others may already have them or be close to having the capability. Most often mentioned are Israel, India, Pakistan, Brazil, Argentina, South Africa and Iraq - states that have not yet signed the Non- Proliferation Treaty (NPT).

It is now widely believed that a total of 14 states, all of them non-signatories of the NPT, have ballistic missiles and that some of those missiles have ranges to 1,500 miles and are able to carry nuclear and chemical warheads. The thought of this weaponry falling into the hands of unstable and warlike states helps to explain the continuing interest in ballistic missile defences that persists in some quarters of Europe and the United States, impractical as it may seem to others. However, the recent experience in the Persian Gulf, where the Patriot anti-missile system was used effectively to defend against the SCUD missile, has lent a renewed interest to the question.

2. Conventional Deterrence and Defence

The words deterrence and defence are linked together in Canada's statements of policy, and for good reason. The demonstrated ability to defend oneself is the finest and most frequently demonstrated form of deterrence, and the world is full of examples. Perhaps because of my interest in architecture my favourite example is the forts and towers, especially the Martello towers in Britain and Canada that demonstrated to the early 19th century the willingness and ability of these countries to defend themselves if necessary. Old Fort Henry in Kingston, Ontario is a fine example of a defensive work that served its purpose without ever having been under an attack or launching one.

As I have indicated earlier, deterrent strength does not need to be as great as that of the potential aggressor. In military terms, it is possible to defend a place with fewer forces than the attacker will need. And since Canada and our NATO allies are interested only in defence we have tolerated certain imbalances in favour of the adversary. For many years we have been able to contemplate without fear the knowledge that the Warsaw Pact had armed forces totalling almost 15 million troops compared with NATO's 12 million, or that the Pact had a larger number of nuclear- propelled and nuclear-armed submarines.

And now, the political eruptions of late 1989, such as the breaching of the Berlin Wall on the 9th of November, the flowering of the Velvet Revolution in Czechoslovakia on the 17th of the same month - these and other events set in motion the breakup of the Warsaw Pact as we knew it, but not before the world witnessed an unaccustomed scene within the Pact. In November of 1990, after the two alliances had signed the historic CFE (Conventional Forces in Europe) agreement, the member countries of the Warsaw Pact met to discuss how the arms reductions were to be apportioned between member countries. This time they met, not merely to be told by the Soviets what they must do, but actually to negotiate the reductions to be apportioned.

Then, on the 31st of March, 1991 the bilateral treaties and exchanges between the Soviet Union and the other members of the Warsaw Pact were cancelled and the Warsaw Pact ceased to exist as a single military force - although the Soviets were still hoping that some non-military union might be preserved. But what was there left to preserve, once the weapons and the threats were gone? And besides, several of the countries were already negotiating with the countries of NATO.

One could, perhaps, assume that the Soviets had finally come to the same conclusion as we in the NATO alliance had reached long before: that a degree of deterrence sufficient to assure peaceful borders could be achieved without striving for massive superiority.

Whether these welcome changes of 1989-1991 took place as a result of NATO's policy of reasonable levels of defence - as I believe - or whether they came about through economic and political disintegration in the Soviet Union would make a good subject for school room debates. The salient facts are: that NATO maintained prudent levels of defence, and that a significantly higher level of prosperity flourished in the free democracies of the West; that Communism proved to be a failed system that ruined the economies and political rights in every country it touched; and that the smaller countries of the former Warsaw Pact have lost no time in seeking new political and economic links with the more successful countries of NATO.

At the heart of Canada's policy and that of other peaceable states is the commitment to maintain a capability to deter aggression in any of its forms and, if that should fail, to be seen to be able to defend ourselves convincingly. This capability must be maintained despite the encouraging events of 1989-91, but at the same time recognizing today's lower levels of tension.

Deterrence must, of course, be proportionate to the expected threat. If the threat increases, the deterrent capability must rise to meet it. Similarly, a lower level of threat - such as the one we are experiencing today - must be met by lowered levels of deterrence and defence capability. To do otherwise would be seen to be provocative or, at best, unresponsive. However, it is also important to bear in mind the realities of the forces in question. Even if the Soviet Union carries out all the reductions of main battle tanks that they have offered, they will still possess the largest armoured force in the world. Even if they keep their SLBM carrying submarines in their home ports or the bastion areas of the Arctic Ocean, they will still possess the largest submarine fleet in the world. In answer to a question put to him by the author at the SCONDVA hearings, Mr. Tariq Rauf of the Arms Control Centre confirmed that the Soviets were still producing these powerful submarines at the rate of 7 or 8 new ones each year.[5]

Although the trend in arms reduction is encouraging, it is well to bear in mind that it has not yet progressed very far. Military policy officers frequently remark that they are paid to be pessimists. Here are a few examples of the start lines from which this move must begin. Under the terms of the CFE, both sides will reduce to equal numbers of tanks (20,000), artillery pieces (20,000), helicopters (2,000), armoured combat vehicles (30,000), and combat aircraft (6,800). Source: Canadian Internation Relations Chronicle, Oct-Dec 1990.

[5] SCONDVA February 22, 1990.

Table 4.
Selected Conventional Force Comparisons

	NATO	Warsaw Pact
Total Armed Forces	12,142,200	14,711,400
Naval Aircraft	3,090	1,055
Aircraft carriers	20	4
Submarines, all types	285	377
Ground Forces in Central Europe *	793,000	975,000
Atlantic to the Urals		
Main Battle Tanks	21,900	58,500
Artillery and Mortars	18,100	49,600
Armed Helicopters	1,100	1,515
Bombers & Attack a.c.	3,228	2,795
Interceptors/fighters	1,200	4,240

* Active force members, excluding Reserves, in what is called the NATO Guidelines area: Belgium, Federal Republic of Germany, Luxembourg, Netherlands, German Democratic Republic, Poland, Czechoslovakia.
Source: IISS The Military Balance, 1989-1990

3. Sovereignty

The concept of sovereignty lends itself to a variety of definitions and interpretations, but all seem to agree that it is the sine qua non of statehood. The White Paper of 1987, reflecting many previous statements out of DND, put it in these words:

> *"After the defence of the country itself, there is no issue more important to any nation than the protection of its sovereignty. The ability to exercise effective national sovereignty is the very essence of nationhood. The Canadian Forces have a particularly important, though not exclusive, role to play in this regard. The protection*

and control of our territory are fundamental manifestations of sovereignty. Our determination to participate fully in all collective security arrangements affecting our territory or the air or sea approaches to our country and to contribute significantly to those arrangements is an important affirmation of Canadian sovereignty...

...an important manifestation of sovereignty is the ability to monitor effectively what is happening within areas of Canadian jurisdiction, be it on land, in the air or at sea, including under the ice. But monitoring alone is not sufficient. To exercise effective control there must also be a capability to respond with force against incursions. Such a capability represents both an earnest of the government's intent to maintain sovereignty and a deterrent to potential violators.

The Government will not allow Canadian sovereignty to be diminished in any way. Instead, it is committed to ensuring that the Canadian Forces can operate anywhere within Canadian jurisdictional limits. Our Forces will assist civil authorities in upholding the laws and maintaining the sovereignty of Canada."

This commitment has still to be put into practice. There is, for example, no surveillance for one entire area in the Canadian Arctic, that measures larger than the British Isles. The Canadian Navy has no equipment for operating on the surface or under the surface in that region of the Arctic.

Our sovereignty is our claim to be a country, to control our own territory, to decide what is best for this nation, and to have the ability to maintain authority over the land, sea and airspace claimed by our Canada.

It has been said that sovereignty only becomes an issue when it is challenged. So it is ironic that the most visible of challenges has come, not from an adversary but from a friendly country. At the moment, the United States considers the Northwest Passage to be an international strait, whereas this country considers it to be internal waters. Our claim is supported in part by the fact that for much of the year the distinction between water and land is very slight.

The Canadian Inuit who live there are in the habit of crossing between islands and mainland, on foot or by snowmobile!

The promised Polar 8 icebreaker (so called because it would be capable of smashing through 8 feet of sea ice) was intended to bring a greater Canadian presence to these waters and to lend strength to our demonstrated occupancy of the waters in this part of our country. The same was to be true of that other famous promise, the nuclear-propelled submarine, the only vessel capable of penetrating all our waters in all seasons.

It has also been stated that a further challenge to our sovereignty might have been made by the floating ice islands maintained by the Soviet Union for scientific purposes and which are subject to drifting with the winds and currents across borders. However, this has never been a serious challenge, and the Soviets have an interest in claiming sovereignty over their own "internal waters". Further, the Polar Bridge Expedition of 1988 gave ample evidence that they accepted our borders as we accepted theirs.

It is worth a moment's reflection, however, that national sensitivities over borders are a common cause of disputes and cannot be disregarded lightly. The joke about the Russian border with Russians on both sides of it was a useful reminder of the Soviet habit of attempting to make themselves secure, even at the cost of making others insecure. We can only hope that this phobia of theirs has been laid to rest.

In peacetime the enforcement of Canadian laws throughout our lands is the responsibility of the civil authorities, and this includes legislation and control over navigation and pollution in Canadian waters, the game laws, the once-rich ocean fishery, and the control of air traffic in our airspace. The military role is to provide coercive force when the capabilities of the civil authorities prove to be inadequate.

SCONDVA heard testimony in 1989-90 concerning the difficulties in enforcing Canadian law over the fishery, especially in Atlantic waters. It was learned that there is a total 343 vessels in the civilian fleet or available to it,

Coast Guard	230
Fisheries & Oceans	84
RCMP	9
Navy	20

The participation of naval vessels in fishery patrols occurs when requested by the civil authorities. Each year the Department of Fisheries and Oceans receives from DND:

900 hours free time on the Atlantic
540 hours free time on the Pacific, and
800 additional free hours of multi-task time.
Beyond those hours, DND charges a fee to DFO.

In response to a question by the author Mr. F. Pouliot, Assistant Deputy Minister of DFO, indicated that the co operation of DND in fishery patrols is welcome and much appreciated. However, the withdrawal of the Tracker patrol aircraft in 1990, to coincide with the closure of CFB Summerside had already, in late 1989, begun to force DFO to contract with private operators for inshore patrols of the fishery.

Further, it appears that Canadian patrol and control of its waters requires the involvement of a large number of agencies. In addition to the four I have mentioned, the department of Indian and Northern Affairs maintains vessels; and the department of Energy, Mines and Resources has an active fleet for mapping and sounding purposes.

In spite of all these many organizations Canada still has problems in enforcing its laws - or could it be because of them? Aggressive poachers, actively breaking our fishing laws, have at times been able to get clean away to the safety of American waters while Canadian commanding officers at sea waited in frustration for clearance from all participating agencies before being allowed to take appropriate action.

The RCMP has a similar interest in upholding the law at sea in the face of an increasing problem of illegal entry of drugs.

It has been observed that the United States manage to get by with only two agencies to control their waters: the Coast Guard and the Navy. The British make do with one, the Navy. In Canada, seven navies exist for maritime functions.

The time has come for Canada to consider how best to re-order the powers and resources for surveillance and control of our maritime sovereignty. A more efficient use could be made of the military and civilian resources now available. We could hope for a system that offers faster response to emergencies at sea and is also cost-effective in a time of fiscal restraint.

4. Peacekeeping

Everyone likes peacekeeping. Even the most hardened anti-defence demonstrator in a street rally will concede that peacekeeping is a laudable activity, a suitable role for Canada's armed forces. Peacekeeping is seen to be non-violent and helpful, therefore acceptable and commendable.

Peacekeeping is also important. Conflicts tend to break out in many parts of the world - at any given time there could be up to 100, depending on one's counting method - and any one of these could have the potential to draw in other combatants and become a major war.[6] By separating and calming down the combatants, peacekeeping aims to quell the disturbance before it can become a conflagration again. Peacekeeping is a valuable service to international stability and peace, but not all countries are well positioned to do it, and not all are as good at it as we are.

Canada is well suited to peacekeeping, and Canada is good at it. As a middle power, well versed in military technology and well known to harbour no territorial ambitions, our contributions have been well received. Canadians have served in such far-away places as Lebanon, Kashmir, New Guinea, Cyprus, Zaire (formerly the Belgian Congo), Korea, Namibia, Central America, Afghanistan, and, most recently on the Iraq-Kuwait border and in the Western Sahara. A complete listing may be found in National Defence publications and in Appendix A to this volume. Most of these operations (23 of the 27 in which we have been involved so far) have been mounted under the aegis of the United Nations, and Canada has always been welcome. Indeed, peacekeeping was a Canadian invention, for which the late Lester B.(Mike) Pearson was awarded the Nobel Peace Prize in 1957.

There are criteria governing Canadian participation in these UN operations, and they may be summed up as follows:

1. The principal protagonists agree to the proposed terms for a cessation of hostilities and to Canada's participation;

2. The arrangements, preferably under the United Nations control or aegis, genuinely serve the cause of peace and have prospects of leading to a final political settlement;

---

[6]The Canadian Strategic Forecast 1990, published by the Canadian Institute of Strategic Studies, listed 28 nations involved in conflict with guerillas or other nations in December, 1989.

3. The size and international composition be appropriate to the mandate and not be damaging to Canada's relations with other states;

4. That Canada have adequate resources to meet a request without jeopardizing other commitments;

5. That there be a single identifiable authority to which participants report;

6. That Canada's participation be for a fixed and limited duration;

7. That the Canadian component have independent transport and communications facilities; and

8. That the participation be adequately funded and logistically supported.

Canada's peacekeeping activities have not been performed without some cost. The conditions are sometimes dangerous and unpredictable, which ought not to surprise anyone who recognizes that guerilla activity is frequently involved. By mid-1991, casualties among the 83,000 Canadians on peacekeeping missions since the beginning had added up to 85 dead and more than 100 seriously injured.

The financial cost of peacekeeping has been a vexing problem for the UN; and for the more than 60 countries that have contributed troops the financial burden has sometimes been greater than expected. To maintain the Cyprus force alone costs an estimated $46 Million for each 6 months tour of duty, or $92 Million per year. The country contributing the troops is expected to bear 70% of the cost, but the UN has not been able to reimburse each country its 30% share of the cost of its peacekeeping since 1980. As a result, it is $225 Million in arrears in that account. Canada has been obliged to contribute 95% of costs to ensure that its peacekeeping forces are well equipped and cared for.

The deployment of Canadian Forces for peacekeeping in Cyprus will cost Canada an additional $11.95 million in 1991-92 above the cost of employing them on defence duties in Canada ($22.5

million). This is something that Canadians might well bear in mind when comparing the costs of defence and the contributions nations make toward collective security. Some of our NATO partners (for example, France and the Netherlands) have participated in only a few United Nations or other peacekeeping missions.

Some peacekeeping operations seem destined to continue endlessly. A case in point would be the United Nations Force in Cyprus (UNFICYP) to which we have contributed since 1964. It is our largest such commitment and, since the fall of 1987 has comprised 576 Canadian soldiers at any given time. The total peacekeeping commitment involves some 1200 personnel, with another 900 on standby in Canada.

Another long term commitment is the UN Interim Force in Lebanon, which dates from 1978. This is a subject that concerns, and ought to concern, the UN Special Committee on Peacekeeping Operations which was created by the General Assembly in 1965 (the Committee of 34). Canada is an active participant in that committee, and through it has attempted to improve the practical implementation of peacekeeping. But Canadian representatives have experienced some frustration with the committee's inability to make decisions. As these problems appear to have been rooted in East - West distrust, we may now hope that the times are becoming more ripe for a new attempt to improve these operations, making adequate provision for rotation of forces and attempting to solve the problems of paying for the operations.

Canadian contingents are frequently requested to provide the communication, logistics, aviation and other specialist units for the entire force, as our forces have a reputation for competence in these fields. And that is a consideration that deserves more than a moment's reflection.

The troops we send to Cyprus, Kuwait and other areas that need us are well trained and suited to the tasks required. Most have been infantry soldiers or members of the logistics, maintenance and communications trades. Most are regular members of Canada's armed forces. Some Reserve personnel have been used in recent years and they have served capably and well; and there is now in place a plan to integrate the Reserves into the ongoing peacekeeping requirement.

Any peacekeeping force able to do the task must be well trained

Canadian Peacekeeper and friend.

for this very demanding - and risky - type of operation. Canadians have proven themselves good at it because they were already good soldiers. Peacekeeping, if it is to be of any real value, must involve not just citizens who are peace-loving, but trained soldiers who know how to go about the job and perform it well.

It should also be appreciated that Canada benefits from peacekeeping in several ways. For one, by helping to stamp out small "brush fire" conflicts far from home we ensure they will not flare up into larger wars that could engulf us; second, Canadian troops gain a significant training benefit from the experience; and third, those same troops gain an important understanding of the broader world in which Canada operates, and an expanded appreciation of the responsibilities of their Canadian citizenship.

The peacekeeping role is one of which Canadians may feel justly proud. It is one of our very tangible contributions to world peace and to the efforts of the United Nations Organization. And it works. Of course, in our pride in this accomplishment, we must never presume to think that peacekeeping is the only answer to global stability and peace. Those other determined efforts we make through collective security, attempting to maintain the right amount of deterrent capability, are the principal ways in which we and our friends attempt to prevent war breaking out in the first place.

With the new found collegiality in the Security Council the potential for more and improved peacekeeping operations seems very good. The Committee of 34 has, over the past two years or more, tendered many ideas for improvements in peacekeeping; and these are being diligently worked on by the Secretariat and the military advisory staff of the Secretary General. One idea that has recently re-surfaced is for a decentralized but standing UN force, ready on short notice to be deployed to potential trouble spots. This force, which would have a permanently staffed Headquarters as part of the Secretary General's military staff, would act as a quick reaction and deterrent force. Its task would be to deploy, where possible before an act of aggression takes place, thus providing a UN "tripwire" to any would-be aggressor. While this force would be armed and able to defend itself, its main role would be as a deterrent, demonstrating UN resolve to provide collective security through quick military action if needed.

In his testimony to the Special Committee of the Senate Lieutenant General R. Evraire added these thoughts, which I think are worth including here:

> *The purpose of peacekeeping is to enable the parties in a dispute to disengage and to give them confidence that their differences can be settled through negotiation... Peacekeeping is a device to assist peacemaking. In other words, the peacekeepers role is to maintain a ceasefire while politicians and diplomats attempt to arrive at a political settlement. In a recent publication Mr. Brian Urquhart, a renowned British peacekeeper, postulated a third aspect which he called peacebuilding, or the effort to rebuild the socio- economic structure of the country. Thus peacekeeping, peacemaking and peacebuilding are interrelated.*[7]

This country should continue its fine work in peacekeeping, meeting every legitimate request within reason.

---

[7] Special Committee of the Senate on National Defence, Issue No. 18 March 15, 1988.

The Canadian government should press the UN Committee of 34 to find a more satisfactory way of apportioning costs for peacekeeping. A general levy on all member countries would be worth considering.

The Canadian government should press for agreement on a system of rotation of troops whenever a new mission is mounted.

Some thought should be given to increasing the proportion of Reserves employed in peacekeeping operations.

5. Arms Control and Disarmament

One of the important pillars of Canadian defence policy is the pursuit of effective and verifiable arms control and disarmament agreements. There has been a broad measure of political support for the idea that security should be obtained at the lowest level of arms possible. Now, it might seem to some that this is not a matter that affects Canada directly, since this country has very few arms left to trade away. For the past 20 years or more, successive governments have reduced the defence share of government spending, in order to divert funds to other programs, with the result that Canada today is one of the more lightly armed countries of the western world.

However, it is in Canada's interest to promote meaningful arms control measures, hand in glove with measures for collective security and defence. Pursued as complementary goals, these twin thrusts can lead us and other nations to the common goal of enhancing security at the lowest level of forces. I know that for some it may seem ironic that members of "the honorable profession of arms" should sit down with diplomats to negotiate the control or reduction of those arms. But who else is better qualified to understand the significance of the equipment and the forces to be discussed? And who has a stronger interest in peace? I have long been persuaded that Canada's armed forces are this country's biggest and finest peace group, and its best motivated.

A publication of the department of National Defence contains some useful remarks on the nature of arms control that are worth repeating here:

The term "arms control" did not enter the strategic vocabulary until the 1960s, although earlier agreements clearly fall into this category. Arms control is a process by which the development, production and deployment of weapons systems and military forces are kept within defined limits according to agreements between states, or the process through which agreements are made that certain weapons will not be used.

Arms control is also concerned with the safe management of crisis situations and the reduction of accidents.

Arms control does not seek to outlaw war, desirable as that may seem, nor does it make the assumption that weapons in themselves are the cause of wars. Instead arms control seeks to reduce the risk of war and, if that should fail, to reduce its destructive effects.

Many arms control efforts have been aimed at making war less inhumane. These efforts have taken on greater importance since the First and Second World Wars and particularly since the invention of nuclear weapons.

Arms control does not seek to abolish military systems, recognizing that states will continue to give priority to their own security, based on self-defence. States that believe their security is threatened or might be threatened will be reluctant to give up their armed forces.[8]

The Canadian Institute of International Affairs has also observed that:

Coincident with the need to maintain deterrent strength is the obligation to join with other peace-loving nations in an effort to reduce the mounting burden of defence and the risk of miscalculation by seeking effective measures of arms control and disarmament. These cannot be a substitute for deterrence and defence but they can reinforce security and stability if they are balanced and verifiable. Negotiations for this purpose take place bilaterally,

[8]Facts About Arms Control, DND, DG Info, 1-88.

between the two superpowers, and multilaterally, between NATO and the Warsaw Pact, in the CSCE and in the UN.

In matters of arms control and disarmament, as in many other matters, the two departments, Defence and External Affairs, work closely together.

Canada's Participation

It pleases me to recall that, even if the term "arms control" is recent, Canada's involvement in the process is not. As far back as 1817 the Rush-Bagot treaty was signed on our behalf, with the United States, to set limits on naval power on the Great Lakes following the war of 1812-14.

Today this country holds the following six specific arms control goals:
- negotiated radical reductions in nuclear forces and the enhancement of strategic stability,
- maintenance and strengthening of the nuclear non-proliferation regime,
- negotiation of a global chemical weapons ban,
- support for a comprehensive test ban treaty,
- prevention of an arms race in outer space,
- the building of confidence sufficient to facilitate the
- reduction of military forces in Europe and elswhere.

Among the better-known arms control fora, Canada has participated in the Mutual and Balanced Force Reduction talks (MBFR) at Vienna, which attempted to agree on troop strength reductions, but was frustrated by lack of ability to agree on each other's present levels of strength; and then at the recently- concluded, and successful, Negotiations on Conventional Forces in Europe(CFE) in Vienna; at the Conference on Security and Cooperation in Europe (CSCE); and the Conference on Confidence-and Security-Building Measures and Disarmament in Europe (CCSBMDE) in

Stockholm. In all of these we are included because of our membership in NATO.[9]

We also take part in the Conference on Disarmament of the UN at Geneva(CD), and in many other fora, including notably those concerned with chemical and biological warfare. Canada does not participate directly in Strategic Arms Reduction Talks, which are bilateral between the USA and the USSR; but we are consulted and kept informed, as are our other NATO allies.

Canada is represented in the meetings of 19 nations of the UN Group of Government Experts on International Arms Transfers. The mandate of that group is to study "ways and means of promoting transparency in international transfers of conventional arms on a universal and non-discriminatory basis". The problem faced by the UN is that of illicit arms transfers, a fallout from the reductions of tanks and other weapons in Europe. The Secretary General of the UN is to report on this subject to the General Assembly in the fall of 1991.[10]

The superpowers began adapting satellites - "national technical means" (NTM) - in the 1970s to monitor compliance with their agreements for limiting nuclear arms. There were at the time no agreed concepts about verifying compliance with those agreements. Canada's departments of National Defence and External Affairs began a broad program to clarify the subject by publishing in 1980-81 a trilogy of studies that are still used as references in international negotiations. However, concepts about enforcing compliance when violations have been detected have yet to be developed.

In 1983 Canada gave itself a distinctive role in arms control when it set up a verification research program under the auspices of External Affairs. This very laudable and impressive initiative has provided yet another reason for Canada being included in arms control consultations, even when its own forces may not be under discussion.

In February of 1990 Canada hosted the first of two Open Skies conferences, intended to produce measures that would increase mutual trust. Just before the Ottawa conference a Canadian military crew sucessfully made a trial flight over Hungary to test the proposal that regular flights could be made by Warsaw Pact and NATO forces over each others' territories to observe each others' military forces. However, the agreement has not yet become a reality.

The idea of Open Skies, which was first floated in 1955 by

[9]CIIPS publishes each year an excellent "Guide to Canadian Policies on Arms Control, Disarmament, Defence and Conflict Resolution."
[10]The Disarmament Bulletin, External Affairs, Spring 1990 p.15.

President Dwight Eisenhower as a confidence-building measure, was always more popular with the Americans than with the Soviets. Fans of Marshall McLuhan might analyse the stand-off as an inevitable result of the cultural differences between the visually-oriented society (the West) and the oral traditions of the East (for whom spying by microphone is more socially acceptable than visual observation.) Others have maintained that it is more a question of the Americans having better equipment for visual observation than the Soviets - a case of the foxes wanting to ban long legs and the rabbits voting to outlaw teeth.

Arms Control Progress

The success of arms control negotiations tends to be a matter of opinion, varying with the start point of the opinionee. Those who favour simplistic solutions and the idealism of "pure disarmament" - the sort who wonder why we cannot all just lay down our arms and set a good example for others - tend to be impatient, citing the long intervals between agreements, and the large number of weapons, especially nuclear warheads, that still exist. Others take heart in detecting a trend toward lower limits on arms, the possibility of real reductions and the usefulness of confidence building measures. The author is in the latter camp, and these few selected highlights of the arms control scene in the past two decades will illustrate why:

1972   SALT I agreement and the ABM agreement signed

1979   SALT II agreement (not ratified by the US Senate but followed in practice by both superpowers for many years)

1979   INF crisis over continuing Soviet deployment of SS-20

1983   INF crisis continues; NATO deploys Cruise and Pershing II

1986   Stockholm conference ends in success: 35 countries agree to security and confidence building measures (CSBMs) to exchange information on troop moves thus reducing surprise and risk of accidental conflict

1987 INF agreement between USA and USSR - to eliminate all inter-mediate range nuclear missiles, ballistic or cruise, with ranges of 500 to 5,500 km and to allow on-site inspections.This was the first arms reduction agreement between the alliances.

1990 CFE agreement limiting Conventional Forces in Europe was signed by 22 NATO and Warsaw Treaty Organization members, with provision for sharp reductions in some categories,such as tanks and artillery, with equality between the WTO and NATO. The parties also signed a Joint Declaration that "they are no longer adversaries"

The Stockholm agreement was a significant breakthrough. It included provisions for notification of military activities involving 13,000 troops or 300 tanks, for mandatory invitation of observers to activities involving 17,000 troops or more,and for surprise inspections on 48 hours notice. One of the purposes of arms control is to make war less likely through some degree of predictability, and the Stockholm agreement did that.

The impact of the INF Treaty was great, and it promises to continue to be so. For the first time the great powers demonstrated that they could not only negotiate an arms control treaty but one that involved the complete elimination of a class of weapon. At the time, some critics scoffed that the warheads reduced would be only some 4% of the total in existence; but others wisely and correctly prophesied that the treaty would represent an important breakthrough, to set in motion other agreements and other reductions. Some very interesting possibilities have emerged through the various proposals that have surfaced since 1987, especially in 1989 and 1990, some of them offering unilateral reductions.

In the summer of 1990 the NATO summit offer of a non-aggression pact with the countries of the WTO and its de-emphasis of the role of nuclear weapons met a prompt and favourable response from the Soviet Union.

Mr. Gorbachev had already offered to make reductions of troops and tanks from Central Europe. At first, some doubted his sincerity but later events indicated that he had very little choice. Poland wanted the Soviet troops out, Czechoslovakia and Hungary wanted them out, and so did East Germany, as it was called prior to the reunification of the two Germanys on the 12th of September.

Some of the reductions agreed under the CFE, represent reductions that are asymmetrical, involving larger reductions by the Warsaw Pact than by their NATO counterparts. But that was only to be expected, considering the preponderance of conventional weaponry of an offensive nature on the WTO side. Thus, for example, the WTO has agreed to reduce its main battle tanks in Central Europe from 58,500 to 20,000 while NATO's reduction to 20,000 is from the present 21,900.

The CFE agreement, signed in Paris on the 19th of November, 1990, is an event of historic importance. Not only did the parties agree to asymmetric reductions but they also agreed that they were no longer adversaries and would extend to each other the hand of friendship.

It has been suggested by some that, to offset these imbalanced reductions by the Warsaw Treaty Organization, NATO should offer something in return. True, but we must analyse it very carefully. The Warsaw Treaty Organization no longer exists except in the minds of dreamers in Moscow. And after all, it is the Soviet Union's huge arsenal of tanks, bridge layers and other offensive weapons that created the original asymmetry, and now the reductions must follow a matching asymmetry.

Something we can offer in return would be our assurances. The communist leaders in the USSR have always been strong for declaratory statements (such as "no first use of nuclear weapons") so Canada and our NATO allies can give some of those in return. We have already begun to do this, by the Joint Declaration of Paris, and the Charter of Paris For a New Europe, signed on the same occasion in November of 1990. Besides, we have already, many times over, declared that our forces are purely defensive, that NATO forces will never be used except in response to attack.

In the summer of 1990 I had the feeling that this was the time to bring those statements out, couch them in the most important, reassuring language possible, and use them as honest payment for the reductions we have a right to expect. NATO and the Warsaw Pact might should enter into an agreement of mutual non-aggression, as proposed at the NATO summit in July of 1990. And that, of course, is what has since occurred. The Soviet leadership may harbour bothersome memories of the 1939 non-aggression pact between Hitler's Germany and Joseph Stalin, but that is their business. Today they have more reason to put their faith in the NATO democracies from which they wish to learn the ways of peace and prosperity than ever their predecessors had to trust the Nazi.

Mr. Gorbachev lost much of his bargaining power in 1990. Between challenges to his leadership from both the right and the left, the crumbling of the Warsaw Pact, and the restlessness of the republics of Georgia, Moldavia, and especially those in the Baltic, he was in no position to dictate terms, and in 1990, in order to avoid his worst scenario - a united Germany free, neutral and uncommitted to the restraints of either alliance, he was obliged to accept a united Germany belonging to NATO. But, as Chancellor Helmut Kohl and his FRG government have offered, there will be no outside troops in the Eastern portion.

Certain other arms control proposals arise from time to time, usually from non-governmental sources. They were pushed with more vigour, it seems to me, during the years when Soviet foreign policy was more aggressive. Some of the more popular ideas raised for discussion in Canada were:

Nuclear Weapons-Free Zone
- Although it is official Canadian policy to support NWFZs whenever they are considered feasible and likely to promote stability in an area they were never a good idea for Canada. For one thing, they were one-sided proposals (the Soviets refused to allow NWFZs at home) and for another, they were unenforcable, even in the open society countries that adopted them. Perhaps their biggest drawback was that they offered people the delusion that some useful action had been taken when nothing of the sort had been done, and this deterred activists from pressing on to some more worthwhile achievements.

Bans on Visits by Nuclear-Powered Ships
- This proposal presumed a confident judgment of absolute condemnation of nuclear power, frequently made by activists whose breakfast toast that morning might well had been made by nuclear-powered electricity. The idea has never been more than an irritant, and made no constructive contribution toward arms control or disarmament.

Bans on Cruise Missile Tests
- This is another one-sided notion, since no comparable pressure was put on the USSR to limit or stop its tests of cruise missiles (and, prior to 1988, the Soviets possessed more cruise missiles than the Americans).

Demilitarization of the Arctic
- This idea was put out by Mr. Gorbachev himself. It was too self-serving to be taken seriously, as it would create a sterile zone in our part of the arctic but not in the part of his arctic that included the many army, naval and air bases in Murmansk and other parts of the Kola Peninsula. It is ironic that General Secretary Gorbachev should have chosen Murmansk as the place to make his proposals for northern demilitarization - the centre of the only significant and continuing concentration of naval and military forces north of the Arctic Circle. It became doubly ironic in view of the later decision to move Soviet nuclear tests to Novaya Zemlya in the Soviet Arctic. Perhaps the idea could be re-worked, but only in a much fairer way. The better remedy lies in a continued reduction in East-West tensions which could lead to meaningful and verifiable reductions in both nuclear and conventional arms.

The Comprehensive Test Ban
- Like the Non-Proliferation Treaty, this is actually Canadian policy, but unlike the NPT, CTB has not yet been adopted and is still stalled in the UN Conference on Disarmament. It is worth noting that testing of nuclear devices has continued, albeit at a slower pace, since 1945.

Table 5.
Worldwide Known Nuclear Tests

	USA	USSR	UK	France	China	Total
1989	11	7	1	8	0	27
TOTAL	921	624	42	180	34	1,801

Source: Ottawa Citizen April 9, 1990

Canada has been a strong and active supporter of arms control for a safer and more peaceful world. We have expertise in verification, including seismic and satellite monitoring; we have access and credibility through our membership in NATO, our contributions through the UN and our acceptability as peacekeepers and observers.

> Arms Control negotiating and monitoring are valuable, honourable ways of contributing to a peaceful world. Canada should continue to make these important contributions.
>
> Canada should also provide leadership for the idea of a NATO Verification Centre as recommended by Mr. Clark in a speech delivered May 26, 1990.

IV FUTURE ROLES, COMMITMENTS AND EQUIPMENT

A. The Constants

Canada has a number of security interests that are constant., and the first group of these can be listed broadly under the heading of the defence of Canada. More specifically, they embrace the surveillance and control of our shores and airspace against military incursion, including the protection of our Arctic.

The second group of interests embraces assistance to the civil authorities as defined in the National Defence Act, including:

Fishery surveillance
Pollution surveillance
Drug traffic interdiction
Surveillance for illegal immigration
Search and Rescue
Disaster relief

A third group of security interests can be called Aid to the Civil Power and is the subject of Canada's Emergency Response laws. For example, units of the Canadian Forces were employed at Oka and other areas in Quebec at the request of the provincial authorities, to end the tensions of those confrontations.

The fourth grouping of interests, I contend, is Canada's share in contributing to world stability and peace - primarily through collective global and regional security arrangements such as those with NATO and the United Nations.

A fifth concern is to safeguard our sovereignty in a continent dominated by the United States. Related to this is a parallel concern that this country not fall into the pitfall of abandoning Europe in favour of defending a "fortress North America". In February of 1990 the author was at a meeting in Brussels attended by Marshall Akhromeyev and General Lobov, and the General said to us that, had there been an American and a Canadian presence in Europe following the First World War there never would have been a Second World War. I believe the remark has much meaning for us today. Further, a German observer has cautioned that:

> *"Together with the United States, Canada forms the North American pillar of the bridge spanning the Atlantic. Because this pillar is composed of two nations and not just one superpower, the latent tension between the European and American pillars is decreased - a valuable contribution to Alliance solidarity that Canadians may not recognize."* [1]

Our ability to cope with these constant challenges has, of course, varied over time. It is the variability - and the credibility of that capability - that prompts the recommendations that follow in these pages.

B. The Variables

In recent years the variable interests have included

1. the nature of the challenge to our security and sovereignty from other countries, in which some of the risks have been reduced, e.g.:

   massive land battles in Europe
   Intercontinental Ballistic Missile attack
   Territorial incursions in Central Europe

2. the nature of the changing international environment in which certain other risks have risen, such as:

   incursions into our arctic waters
   illegal drug trafficking
   illegal immigration (including trafficking)
   illegal fishing in Canadian waters
   international terrorism, including, for example:

   • IRA attacks, sometimes on innocent parties,

   • Muslim fanaticism, exhibited in the extreme by the fury over the writings of Salmon Rushdie

[1] Heinz von zur Gathen, MacKenzie Paper no. 10, p.14.

- The PLO and its continuing reign of terror directed against civil aircraft, cruise ships and other convenient but innocent targets

3. the need for enhanced peacekeeping activity:

There have been Canadian contributions from the beginning: the UN Observer Group in Kashmir in 1949, the Truce Supervision Commission in Indochina (now Viet Nam, Laos and Cambodia) in 1954, peacekeeping operations in the Middle East in 1956, and now recently in Afghanistan, Namibia, Central America, the Western Sahara and Kuwait. A related tasking arose with the need to train several thousand Afghans in techniques of recognizing and clearing explosive mines, and providing assistance after Hurricane Hugo in the Carribean. And by early 1991 Canadians were in Kuwait assisting with mine clearance. Three hundred of our trained troops have been committed to the Persian Gulf area for these duties. Each year brings new challenges, and Canada is a favourite country for the UN to approach for peacekeeping forces.

C. Considerations and Priorities for Defence

NATO is re-aligning its roles and Canada, too, has an opportunity now to re-assess its own defence roles and emphasis. The modest level of our contribution to the NATO effort has for many years been an embarrassment to some Canadians, too much for others and, apparently, a matter of indifference to many. Are 20 ships enough? Are 77 tanks - 25 years old - appropriate in Europe? Do we need three squadrons of aircraft in Europe? Now is as good a time as any to sort out how we may best match our contributions to the needs of the international community. Europe is a good place to start.

As the authors of the White Paper observed, Canada has deep roots in Europe and we care about its fate. It would be hard to imagine a world without a free Europe, and we would not care to contemplate it. Besides, a Europe at peace tends to result in a Canada at peace. So for all these reasons it is in our interests to retain our European ties, especially those we have through NATO.

But is it important or necessary for Canada to maintain the same forces in Europe? Consider, for example, the personnel we now have there. There is the 4th Canadian Mechanized Brigade Group (4 CMBG) stationed at Lahr, Germany, and the 1st Canadian Air Division at Baden-Soellingen (1st CAD), including two Wings, an air defence regiment and support groups. Personnel (in 1989) consisted of approximately 8,000 military and 4,400 civilian employees plus dependents, for a total of about 21,500 Canadians in the Black Forest region of West Germany.[2] A decision has already been made to reduce these numbers, and by the 1st of May, 1991 the military personnel were down to about 7,500.

The two most conspicuous (and most expensive) items of equipment operated by these Canadians are the three squadrons of 44 CF-18 fighter aircraft at Baden and the 77 Leopard I tanks of the armoured units at Lahr. These tanks are 1966 technology. They are now obsolete and no longer able to compete successfully against the tanks and other weaponry of the 1990s. And they are very few in number, by the standards of the European continent. Consider, for a moment, that the combined nations of (what was) the Warsaw Pact have 58,500 in Europe and that NATO countries have an estimated 22,000 in the same space, as reported in Chapter III.

Consider also, for instance, the West Germans have 5,000 main battle tanks, the British 1,300, the Italians 1,720 and the Dutch 913, almost half of the latter being modern state-of-the-art Leopard IIs. Alongside these, Canadian Forces in Europe can mount a total of 77 obsolete Leopard Is.[3] The numbers invite humorous and unflattering comparisons.

It is a Canadian failing to spread ourselves too thinly. We do it in foreign aid programs, and we do it in defence.

Since the tanks are due for replacement, now is an excellent time to make an important decision. The tank is a weapon designed for the massive land battles for which Europe, unfortunately, became famous in two world wars, as did North Africa and the Middle East. It is of limited use in Canada, and not worth having here except as a tank trainer.

The major powers have traditionally striven to maintain balanced forces to meet their security needs, and this includes balanced armies having infantry, armour, artillery and support. Infantry and armour have gone into battle together when available, and many veteran soldiers will wince at the thought of doing without tanks. But, unless Canada intends to expand its land forces to

[2] Defence 89
[3] IISS The Military Balance 1989-1990.

A Leopard tank of the 8th Canadian Hussars crosses a pontoon bridge during a NATO exercise in the Federal Republic of Germany.

enable them to operate overseas at divisional strength, with Brigade strength additionally available for contingencies, it is difficult to see why our tanks ought to be seen to be important. They are expensive to buy and may have little value in the quantities we are likely to place in the field. It may be that Canada's most useful contribution in Europe will be as rapid reinforcements, and as rear-area defence. If that is the case, then most of Canada's foreseeable army needs will call for only light infantry and light artillery, with vehicles to match.

For many other countries geography shows the way to the type of armed forces required. The Swiss, of course, have no need for maritime forces, having no maritime areas. The Swedes, being small and neutral, want only to defend their own land, sea and air spaces. Nineteenth century empires, such as Victorian Britain, attempted to cover all contingencies. It was said that the British Navy was the gun that propelled the Army (the projectile) to where it was needed. In recent times the forces of the United States have resembled that model, at least in concept, and the

Soviet forces have attempted to do the same despite being primarily a land power. Admiral S.G.Gorshkov's legacy to his country was its blue water navy.

For Canada the defence problem is less clear than that of, say, Sweden, and less specific. Instead of arming to protect our own territory - which is too vast and inhospitable for easy defence - we direct our efforts instead to preserving world stability and peace, so that we may never suffer the consequences of a severe degradation of that peace. The closest we come to adopting the defence posture of our larger allies is in our determination to protect the sea lines of communication. Equipment chosen for Canadian forces should be of a type suitable for several options and not too specialized. It should first fit into the Canadian context of surveillance and deterrence against aggression into our land, sea and air spaces.

The Chief of Defence Staff addressed the Standing Committee on National Defence (SCONDVA) in late 1989 and, putting the case for a defence capability beyond the minimum of home defence and peacekeeping he said:

> *" The price of being a sovereign nation is having the ability to defend yourself or to play your part in alliances for defence. It is also the ability to participate in international attempts to stem violence... I foresee a requirement for forces that can participate in whatever defence of this country is needed, given whatever threat exists. These forces will assist our allies, be they multilateral allies or bilateral allies, in assuring not only our defence but theirs. These forces will provide a capability for projecting Canada's will beyond our borders. Also they would assist in peacekeeping operations, or indeed other measures for verification (of arms control)"*[4]

We must maintain the ability to defend ourselves - in co-operation with others; we could never do it alone - and in the process to contribute to making the world more stable and peaceful. But now, in these times of reduced tension and reduced budgetary expectations, it is time to be more selective in our defence efforts, concentrating on what we do best and on what is most useful to those ends. In summary, it is suggested that the external tasks for the

[4]General John de Chastelain, at SCONDVA 19 Oct 1989.

Armed Forces should concentrate on:

a. Peacekeeping,
b. Maritime escort and surveillance,
c. Air transport,

and the domestic tasks should concentrate on:

a. Safeguarding our sovereignty,
b. Surveillance and control of our shores and airspace against military incursions,
c. Aid to the civil power, for maintenance of law and order when requested,
d. Assistance to the civil authorities
   • Search and Rescue
   • Fisheries surveillance
   • Pollution surveillance
   • Drug traffic interdiction
   • Illegal immigration control
   • Disaster relief

I shall not presume to calculate the exact costs, reductions or increases to be expected, but I have long recommended that Canada's defence budget be increased by the year 2000 to 3% of GNP.

National Defence is Canada's most important peace group

1. Maritime Forces

In defence considerations Canada is a coastal state. With our 244,000 kilometers of coastline, including the islands of the Arctic, Canada has the largest coastline in the world. Of every dollar generated in the country 35% is generated by international trade, and 55% of that trade goes to its final destination across some body of water. Nearly 61% of this country's oil and gas discoveries made in the year 1985 were from the offshore. And, some years ago Canada "elected to supervise and control a very large (area of) oceans, of water which (amounts to an area equal to)

60% of the land area of Canada."[5] And yet, as one of my parliamentary colleagues has remarked, we may be "...the only country in NATO that is not capable of defending its own coastal waters but can somehow find $1.2 billion a year to send a token force to Europe."[6]

We have a small navy, consisting of under 10,000 regular force personnel, 4,200 reservists, 18 fighting ships (destroyers and frigates, three aged submarines, support vessels and a small number of aircraft assigned by Air Command. The listing below was provided by the Chief of Defence Staff in November, 1990.

Table 6.

MARITIME FORCES

Personnel
Regular	10,000
Primary Reserve	4,200

Major Operational Units
Destroyer Squadrons	4
Submarine Squadrons	4

Principal Equipment

	East Coast	West Coast
Frigates/Destroyers	10	8
Reserve Frigates	0	0
Submarines	3	0
Replenishment Ships	2	1
Long Range Patrol Aircraft*	14	4
Helicopters(Sea King)	31	4
Diving Support	1	0
Training Vessels	21	0

Bases in Canada	3

*operational control only(Air Command resource)

---

[5] Capt. L. Hutchins (ret) and Capt. P. Godbout (ret) of the Naval Officers Association of Canada, at SCONDVA Dec. 12, 1989.
[6] Mr. Derek Blackburn, MP for Brant, at SCONDVA June 1, 1989.

I think Mr. Alex Morrison, the Executive Director of the Canadian Institite of Strategic Studies, has best summed up the maritime need in these terms:

> *With regard to Maritime Command, we ought to have resources capable, first of all, of telling us to a reasonable extent what is happening in our three oceans. Second, we ought to be able to support our commitments to the North Atlantic Treaty Organization, which I gather are supported by well over three-quarters of the people of Canada, and we ought to have an interdiction capability; that is, we ought to be able to seek out those who are violating international agreements reached with the Government of Canada or those whose indications are such that they might be violating international agreements reached with the Government of Canada.*

In adding the following he provided a capsule summary of what the navy does today:

> *I see the navy as an element of a Canadian defence force that is able to carry out what we might call non-traditional military tasks, but that is capable of performing a general combat role. The thing that must come first with the navy is the equipment and the training of the personnel in a military sense. Once they are able to do that, they can be deployed on fisheries tasks, on drug tasks, on refugee tasks, on sovereignty tasks, but all the while they retain the capability to engage in general combat. I think any armed force worthy of the name must first of all possess that capability and then, of course, because of the training received and the equipment provided, be able to carry out the other tasks.*[7]

It is useful also to repeat here the words of the Naval Officers Association from their appearance at SCONDVA:

---

[7]SCONDVA January 25, 1990 p. 18.

We conclude that life on this planet is intimately dependent on the continued health of the maritime environment, that the oceans are reaching a saturation point where they can no longer continue to absorb the unabated release of pollutants from any source, and that these pollutants are even now damaging ocean ecology.

We conclude that the exploitation and use of the ocean's living resources are crucial to the economic and social survival of the world's people and that fishing disputes between countries and foreign fishing interests can have consequences far beyond the specific issues of fishing. We conclude that the exploitation and use of the ocean's non-living resources are becoming increasingly important to a number of countries and to corporate mining interests.

The world's economy is becoming increasingly interdependent, and this has elevated the value of and role played by international sea-borne trade. Given the nature and impact of modern technology on the evolution of sea-borne weapons systems, the oceans are playing a major role in the international and national defence and security.

The first of Canada's new frigates. HMCS Halifax during sea trials in 1991

These witnesses also reminded the Committee that, of the $80 billion spent annually on Canada's ocean activities, roughly 30% is spent in the fisheries sector and 25% in offshore oil and gas exploration. Nonetheless there have been some puzzling oversights in our attention to ocean problems and in some of the things we do not know about our shores. For example, there is no published map of the ocean floor off Newfoundland, and yet we are drilling holes in the bottom there, looking for oil.[8]

Canadians ought to be concerned about the state of maritime defence and related activities. The extent to which the navy can perform its job at present owes much to the fine quality of the officers, men and women who serve it. Their professionalism and devotion to their tasks has long been a point of pride. But the same cannot be said of the equipment they are expected to use, the ships in which they are expected to live in safety while performing their jobs effectively.

Canada's navy today has only 18 fighting ships (recently reduced from more than 20, since some of the older ships are at last heading for the scrap yard, and the first of the new patrol frigates is not yet in service). In the peacetime navy of the 1960s there were more than 60. Are 18 or 20 enough? Can they defend themselves and the merchant ships they are expected to protect?

The most modern vessels are known as the 280 class. They are only four in number and, as they were built in the 1970s, are now undergoing a midlife refit known as TRUMP.[9] The Saint Laurent class of destroyer dates to the 1950s - well beyond the normal 20-year lifespan of a naval ship. For Canadian sailors and ex-sailors one of the most telling remarks was the one made by a former commanding officer of HMCS Saguenay when invited to the ship's 25th anniversary. He remarked that it was a sad day for such an occasion, as no fighting ship was ever intended to stay in service that long.[10]

A fighting ship is said to have three modes of action: to float, to move, to fight. If the ship is expected only to float and to move, then its age might be unimportant, at least not a cause for concern. An old but serviceable ship can be used for basic training in everything short of fighting. But if a ship is expected to protect shipping and perform other wartime tasks in times of danger, Canadians have a right to ask whether enough resources have been devoted to meet the need. Today's fighting ships need the capability to con-

[8]Hutchins and Godbout of NAOC at SCONDVA, December 12, 1989.
[9]Tribal Class Update and Modernization Program.
[10]G.H. (Skinny) Hayes..

tend with the missiles and other technology of the 1990s.

Canada has had a recent reminder of this need. When the crisis developed in the Persian Gulf in the summer of 1990 Canada's navy was caught without ships capable of meeting the likely hazards. So a hurried refit was given to HMCS Athabaskan, Terra Nova and Protecteur to arm them with the means of defending themselves. That done, the ships proceeded to the Gulf where they performed a very creditable job of enforcing the UN's economic sanctions against Iraq. With only 10% of the ships present, ours performed 25% of the ship interceptions. (But it must be admitted that the reason for this rate is to be found partly in the condition of the ships. Since the Canadian ships - even with the new weaponry - were not capable of frontline duty they were stationed in the rear areas; and those were the areas where merchant shipping was to be encountered.)

The way in which this country has gone about building ships to provide for maritime security is a source of wonderment and dismay. It has been a feast-and-famine situation for many years. During the years when no ships are ordered the dockyards lose the skills and, by the time an order arrives, problems appear in estimating, in fabrication and in many other aspects of the process. There must be a better way.

> Since the number of fighting ships we are keeping at present is approximately 20, and since the life expectancy of such a ship is 20 years, a way should be found to award contracts for the construction of one such ship each year, thus keeping the art alive in the shipyards as well as a predictable and assured supply for the users. Canada should explore the possibility of marketing frigates to allied countries

Having paid the start-up costs for a new shipbuilding program, government and industry will both reap the benefits of this learning by continuing, without interruption, the learning process. Such a move would also allow industry to incorporate technological advances in an orderly fashion.

Frigates are already very expensive and will become more so in the future. One commentator, writing in a defence publication, has suggested that what might make more sense in future is a larger number of small corvettes because they "do not warrant such extensive self-defence systems. The net result is that these combinations (of corvettes and multi-aircraft (twin hull, or SWATH ships) can provide a larger number of sensors and weapons than can equivalent groups of frigates."[11] There would be another advantage to having a navy with a large number of small ships; it would greatly improve the opportunities for personnel to obtain experience of command - a weakness in a navy with few ships, and one not quickly remedied in time of need.

On the other hand, there is a strong body of opinion that maintains the 4,000 ton ship (the new frigate) is the best all-round solution to modern problems, taking into account the modern weaponry required, including large shipborne helicopters.

Speaking of which, the shipborne aircraft - the Sea King helicopters - are overdue for replacement. The New Shipborne Aircraft program should get, and seems to be getting, a priority. The Aurora long range patrol aircraft under the operational control of Maritime Command is a good platform but, with only 18 of them it is difficult for them to perform a credible job of surveillance coverage. One arctic patrol per month is hardly credible. And the three Arcturus craft intended for augmentation of this fleet had not yet entered service in May, 1991.

Our harbours are undefended, vulnerable to minelaying that could immobilize our ability to move ships in an emergency. The two small ships "taken up from trade" in late 1988, HMCS Anticosti and HMCS Moresby, represent a slow beginning toward giving Canada once again a minesweeping and mine hunting capability. They are an economical, interim solution to the problem and offer an opportunity to get the training underway for the Reserves who will crew them. The additional 12 mine countermeasure vessels, now called the Maritime Coastal Defence Vessel (MCDV) progressed in 1990 from the project definition phase to the project design phase of procurement. When these new Minor War Vessels are ready for sea they will benefit not only the security of Canada but also the Reserves, for whom this role is well suited. It will not be too soon.

Then there are the "Trackers" - those noisy, low-speed, low-level, twin-engine patrol aircraft that until recently provided "the sound of security" to the people of Summerside, P.E.I. as they conducted the inshore patrols. Taking these dependable aircraft

[11]Michael Eames, Canadian Defence Quarterly, December 1989.

The Tracker – for many years the inshore patrol and surveillance workhouse of the Canadian Forces.

out of service, instead of giving them the update that had been promised, seems to have been an economy move aimed at closing the Summerside base. The way in which they have been replaced is not completely reassuring. Private contracts have been let to provide unarmed light planes to conduct the close-in coastal patrols, while the long-range Aurora is expected to cover the rest, along with its other, overly-burdened commitments. It would be interesting to see a study of the usefulness of acquiring Canadian-built aircraft such as the DASH-8 for the purpose.

Arctic sovereignty is a continuing concern. The matter first came to public attention in 1969 when the U.S. tanker Manhattan attempted to traverse the Northwest Passage. In 1988 the voyage of the icebreaker Polar Sea attracted similar publicity. The submarines of our British and American allies have been able to surface through open water near the North Pole, presumably having travelled through waters of Canadian interest to get there. Soviet nuclear powered submarines also possess this capability. Canadian vessels do not.

Pollution in our waters is a growing concern. Toxic wastes can be very harmful everywhere, and no more so than in the Arctic. The dumping of all waste, including the pumping of ships' bilges off any of our shores is a source of concern. The use of driftnets in the Pacific is alarming. Illegal fishing in Canadian waters, illegal immigration and drug smuggling - these also form part of the circle of concerns for Canada's defence and protection of sovereignty. Of course, these concerns are not those of the navy alone, many of them being of a non-military nature. Now, as the nations move toward a more peaceful world, attention can and should be directed to those non-military problems in which the navy can be useful to Canadians.

To cope with all of these concerns this country is blessed (or cursed) with a large number of agencies having responsibility and a claim on the public purse in order to meet it. Witnesses to SCONDVA estimated that there are 13 federal agencies with responsibility for some aspect of protection of Canada's oceans. Seven of them have vessels, ships or small boats, for this purpose:

Maritime Command (the Navy)
Coast Guard
Fisheries and Oceans, Canada
Royal Canadian Mounted Police
Hydrographic Service
Indian and Northern Affairs, Canada
Energy, Mines and Resources, Canada

There is compartmentalization of responsibility and there is duplication of effort, of stock control and of vessels that could perform more than one task. At a time when all departments including DND are claiming that their resources are strained, it would appear that there are savings to be found, if a better means of coordination of activities could be found.

There is no one spokesperson for the oceans of Canada, no one Cabinet Minister or Assistant Deputy Minister to co-ordinate and make it ocean policy more efficient and more effective. In the United States, as already mentioned, maritime functions are all performed under two bodies only, the Navy and the Coast Guard. In Britain the Navy does it all. In Canada it is as though we had seven navies.

Whatever else may be done, it is obvious that the navy has an important role to play. The navy has the equipment to a greater extent than the other agencies. But it does not have the powers of a peace officer, such as those possessed by Fisheries officers or

78 THEY STAND ON GUARD

the RCMP. There is a need to have those peace officer powers better consolidated.

SCONDVA also heard testimony concerning the long response times consumed during actions to police illegal fishing and other activities. The Concordia case[12] revealed a certain amount of helplessness suffered by the naval commander on the spot in the face of the long process required in order to obtain permission from a number of sources, before being empowered to take action against the offending vessel.[13] Perhaps it would be instructive to review the operations of our own Search and Rescue coordination centres, for inspiration in solving the maritime problem of coordination.

An important part of Canada's contribution to NATO is to be found in STANAVFORLANT, the Standing Naval Force Atlantic, to give it the proper title. The editor of STARSHELL has described it thus:

Ships pf NATO's Standing Naval Force Atlantic (STANAVFORLANT) replenish at sea off Portugal. Clockwise from centre are: HMCS Protecteur (Canada), HMCS Algonquin (Canada), FGS Emden (Federal Republic of Germany), USS Taylor (USA), and MMS Danae (UK).

---

[12]U.S. fishing vessel in fishing dispute off Nova Scotia in 1989.
[13]SCONDVA March 8, 1990.

"The armed forces of these nations are normally under national command and are only assigned to NATO when needed. However, to provide a continuous NATO presence and a constant and visible reminder of the solidarity of the Alliance, to provide NATO with immediately available deterrent forces, and to have a platform for the improvement of NATO's capabilities and evaluation of tactics, some forces operate under the command of the Major NATO Commanders at all times.[14]

STANAVFORLANT is such a command, and Canada contributes one ship at all times. When this fleet is not at sea, training together or exercising together, it can often be seen paying goodwill visits to foreign ports.

Canada should consider better co-ordination and consolidation of its maritime responsibilities, for greater effectiveness and possible savings in managing its ocean resources. Placing all non-naval responsibilites under one Minister would be one worthwhile way of approaching it.

Participation in STANAVFORANT is beneficial for Canada as it is for NATO and should continue.

For better coordination of assistance to civil authorities a common operations centre, similar to that of the Search and Rescue operation, should be considered.

Canada's navy possesses three Oberon-class submarines. A fourth was bought in 1989 for training purposes, but it will not be capable of leaving the dockyard. The nuclear-powered submarine option was shelved in the April 1989 budget, in spite of well-thought out and very convincing arguments on behalf of the better-balanced fleet that it would represent. The need remains, and if we are not to lose all semblance of ability to perform patrol and surveillance of our sovereign waters, this gap should be filled at once.

On December 13, 1989 SCONDVA produced an interim report on Canada's maritime sovereignty, for the House of Commons. This report observed that "One aspect of Canada's defence which

[14]STARSHELL, the newsletter of the Naval Officers Association.

will remain constant...is the need for its armed forces to maintain a strong sovereignty protection capability...(and) to preserve Canada's security in an era of radical change while identifying the needs of the future." This report also recommended:

> *"In view of the ever-increasing importance of the surveillance of Canada's coastline, the Committee recommends that a strong air surveillance capability by the Department of National Defence be not only maintained, but escalated..."*

and

> *"The Committee ...recommends that the decision to acquire new conventionally-powered submarines, appropriate for Canada's requirements, be made without delay in order to ensure the timely replacement of the Oberons."*

The Committee's final report, which appeared in November, 1990, contained 18 recommendations in all, including:

> *"the installation of fixed accoustic sensors in Arctic waters capable of detecting intrusions...*
>
> *"that research be conducted as a high priority into methods of stopping uncooperative boats on the high seas without endangering human life...*
>
> *"redraft an oceans policy for Canada that takes into account the importance of Canada's Armed Forces in guaranteeing national sovereignty...*
>
> *"that the government appoint an independent panel...to study...government divisions of (maritime)responsibility with a view to rationalization...*
>
> *"a major public review of Canadian security policy ...by 1 Jan 1992*

I support these recommendations and I further recommend the following:

Recognition of the fact that Canada has no maritime capability for establishing surveillance and control of our sea and undersea territories, i.e. for exerting Canadian sovereignty in those areas we claim as Canadians. Maritime aircraft are no substitute for undersea surveillance in ice-filled waters.

A decision should be made immediately to proceed with a program of conventionally-powered submarines to replace the Oberons. The first submarines of a minimum six-boat program should be of the diesel-electric type, but the technology for an air independent propulsion system should be studied and evaluated. If that technology matures over the next five years the Department of National Defence should be ready to incorporate it into the remaining boats and as a retrofit plug into the first subs off the production line.

The government should no discount the possibility of acquiring a nuclear-powered submarine fleet at some future time.

The two batches totalling 12 frigates that are now on stream should be completed and put into service as soon as possible.

The program for acquisition of Maritime Coastal Defence Vessels should proceed as rapidly as possible.

The program for replacement of the Sea King shipborne helicopters with the new EH 101 should proceed as rapidly as possible.

Given the need for a strong air surveillance capability in coastal and open waters, consideration should be given to doubling the number of Aurora or similar long range patrol aircraft fro the present 18, as recommended by the Senate sub-committee on National Defence in 1983.

The Naval Reserves should be increased in size, with particular attentio nto specific tasks, including Naval Control of Shipping and Mine Countermeasures, but without restricting the opportunities for transfers between Regular and Reserve forces.

> *SAILORS, WITH THEIR BUILT-IN SENSE OF ORDER, SERVICE AND DISCIPLINE, SHOULD REALLY BE RUNNING THE WORLD.*
> sign in HMCS Bytown officers' mess

2. Land Forces

Canada is noted for its large land mass, the second largest in the world at 9,976,139 square kilometres. As there has been little in the way of a direct threat to Canadian territory, our armies have been deployed mainly in Europe, where the global front is perceived to be, and in various countries on missions of truce supervision and peacekeeping. Recently we have been gaining awareness of the uses that might be made of land forces in the Arctic, in the preservation of our sovereignty and in demonstrating our occupancy of the territory we claim as ours. We could never hope to defend this vast country by ourselves, but we can and do find security in defensive alliances - NATO and NORAD - and we have developed mobile troops that can quickly move where they are needed.

A logistics officer and a sergeant from 4 Service Battalion, CFB Lahr, discuss tactics during a NATO exercise in West Germany.

In fact, our army is very small in relation to the size and even to the population of the country, with only 22,500 members of the Regular force.[15] Unlike many comparable countries our Reserve force - the Militia - is even smaller at 20,000. The listing below was provided by the Chief of the Defence Staff in November, 1990.

Table 7.
MOBILE COMMAND FORCES

Personnel
Regular	22,500		
Primary Reserve	20,000		

Major Operational Units
Brigade Groups	3
Special Service Force	1
Task Force Headquarters	1
Canadian Division HQ	1
Helicopter Squadrons*	11 (4 Reserve)
Major/Minor Reserve Units	132

Principal Equipment

	in Canada	in Europe	with UN
Main Battle Tanks	37		77
Armoured Vehicles,General Purpose	670	14	6
Armoured Personnel Carriers	614	474	2
Reconnaissance Vehicles (Lynx)	112	60	2
155 mm Artillery	50	26	
105 mm Artillery (towed)	223	10	
Anti-tank Defence(towed)	98	53	
Army Aircraft(Kiowa,Twin Huey,Chinook)	98	0	
Bases in Canada	8		

*operational control only (Air Command resource)

The major formations of the land forces are three Brigade Groups and a Special Service Force, all supported by helicopter squadrons. Land forces are deployed in Canada and Europe, under Mobile Command, which is located at St. Hubert, Quebec.

Traditionally, Canada has relied on its Reserve, or Militia, forces to be mobilized in time of war. This was the case in the two World Wars fought this century. It was only the advent of the

---

[15]Personnel in Mobile Command include 6,500 civilians for a total of 51,150. These and other personnel data may be found in Defence '90.

Cold War, with its uncertainties, that required this country to maintain a large standing army, a major portion of which was either deployed in NATO Europe or designated for quick deployment there.

Then defence spending was cut back over the years, starting in the 1960s, and the army was reduced in size, Regular and Reserve alike. Armouries across Canada, which had been home to militia units and a link to the communities, were closed. Some towns, even those with a long and proud military tradition, towns where men and women had for years participated in the training and in the social life of the militia, have lost this link. There are many towns in Canada today with no military presence. As with most of the defence ministers I have known, I would like to see the day again when every street, in every town, would have at least one young Canadian who wears the uniform on parade nights or for weekend exercises.

Given the events in Europe, marked by the emergence of the European Community and its plans for 1992, by re-unification of Germany and by the democratization of many of the countries of the former Warsaw Pact, how important is it to Canada, to NATO, to global peace, that we continue to maintain the two bases at Lahr and Baden?

Our Canadians in Europe live now in more than 125 communities near the two main bases in the Black Forest region of Southern Germany. To maintain our 20,000 Canadians in Europe, with their equipment and their lifestyles, is costing Canada $1,147,649,000 per year.[16] However, this figure should not be left unexplained. Operating expenditures, consisting of personnel and operations and maintenance costs account for $764,535,000 of that, expenses that would have been incurred even if those troops had been based in Canada, but that would have been spent in Canada.

Two knowledgable observers, Professors J.T.Jockel and Joel Sokolsky have suggested that there are problems in attempting to maintain our present level of capability:

> ...Even before the April cuts the army's plans for CFE were problematic. The peacetime overcrowding at Canadian Forces Bases Lahr and Baden-Sollingen is shocking. Above all, though, 5GBC has to be transported in wartime from Canada. Because there is insufficient air

[16] National Defence 1989-90 Estimates, Part III, p. 51.

> *transport, much of its equipment would still have to go to Germany by sea. Thus the shift in destination for 5GBC from Norway to Germany would not have entirely alleviated the serious transportation difficulties.*
>
> *But that is not all. The division concept was predicated on there being new tanks and other equipment not only for 4CMBG in Germany but also but for 5GBC as well. Now, only one set of new equipment is to be acquired, and even then the most crucial component, the new tanks, is to be delayed.*[17]

What they say is true but, if we were to change the basing concept, if we were to make more use of Reserves, if we were to change the roles for the forces in Europe - to emphasize rear area security and defence, which is more like peacekeeping and needs light equipment instead of tanks - the picture would look more encouraging.

It seems that the defence debate these days involves the Continentalists and the Atlanticists. I want to make it plain that I am not a Continentalist, not a supporter of the notion of "Fortress North America". I am a firm believer in NATO and especially in its transatlantic dimension, but I think that we may be able to serve peace and NATO in new or other ways.

We could begin by asking the question: why are we in Europe? why are we in these bases at Lahr and Baden? The first answer is the one already provided in an earlier page: to exercise together with our allies, to work together, to show solidarity. These reasons are important. So is the reason offered by Bernard Wood, Chief Executive Officer of the Canadian Institute for International Peace and Security"

> *"Canadian participation, through our NATO contingent, has taken on heightened political importance by helping buffer European-US relations during the all-important process of East-West Negotiations and the reductions that will follow."* [18]

But is there another reason as well? Is it really necessary to be there all the time, to maintain schools and skating rinks, housing for spouses and children, chapels and playing fields and all the

---

[17] J.T. Jockel and Joel Sokolsky "Defence White Paper Lives Again" International Perspectives, July/August 1989.

[18] Bernard Wood "Peace in out Time", Director's Annual Statement 1989/90 p.21.

trappings of a Canadian town in another land, including a golf course? Could it be that we just like being there? that we like having these bases to visit and in which we politicians, bureaucrats and business representatives can feel welcome and at home?

Perhaps the time has come to design a new arrangement whereby Canada remains in Europe, and is seen to be remaining there and performing the same useful functions, militarily and politically, as at present. Perhaps the time has come to look again at the North Norway option as an alternative to the Central front, in parallel with another option that has not yet been explored.

First, let us consider the CAST commitment. Prior to June of 1988 this country had a commitment to reinforce North Norway in time of crisis with the deployment of two fighter squadrons and the CAST brigade (Canadian Air-Sea Transportable).

In August and September of 1986 the feasibility of the commitment was tested in what was called Operation Brave Lion, a major exercise this author was privileged to observe. It was judged that the three weeks time it took the forces to get to Norway and ready to operate before an emergency could become a war proved that the CAST commitment was not feasible. Then, in June 1987, the White Paper announced that:

> *"The task of the CAST Brigade Group will be shifted to the central front thus enabling the Canadian army to field a division sized force"* [19]

and

> *" Canada will contribute an Air Division capable of multi-role operations"* [20]

This was the consolidation of forces in Europe. It meant the abandonment of the CAST commitment in Norway together with an expansion of strength in the Black Forest region, centred on the bases at Lahr and Baden. Our commitment to the Allied Mobile Force (Land) would still remain - at the battalion size - and our equipment that we had pre-positioned in the Tromso area would remain.[21]

But then the Budget of April 1989 cast doubt on the tank replacement, on the Division strength and on the expansion of the Reserves, all of which had been factors in the decision to abandon Norway and to consolidate in Central Europe.

---
[19] White Paper p. 62.
[20] Ibid.
[21] Canada has a maintenance contract with Norway for the testing and repairing of our pre-positioned equipment as required.

It is now time to ask whether, in the light of these and other circumstances, the earlier idea should be re-visited, and bearing in mind that:

- the equipment is still pre-positioned in Norway,
- neither Norway nor Denmark will allow permanent stationing of troops on their soil in peacetime,
- tanks make little sense for Canada,
- tanks are mainly useful in the Central front of Europe,
- world tensions have eased, and the warning time now expected by strategists is up to 45 days (not the 21 days that Exercise Brave Lion was allowed).

Allow me to cite again the opinions of Professors Jockel and Sokolsky on the subject

"The army's plans for Europe are now clearly in complete disarray. As General de Chastelain told the Senate Committee, the "plan to field the combined force... without new equipment, entails some risk." That is an understatement. Canadian soldiers often referred to their Norwegian commitment as "Hong Kong in the snow, "recalling the disaster that overcame an ill-prepared Canadian unit in that British Crown Colony in December 1941. Now, in the event of another great war, Canada's soldiers would be facing a Hong Kong in the Black Forest.

Back to Norway
It has often been overlooked, in the talk about Canada's 1987 "recommitment" to the Central front and "abandonment" of the Norwegians, that the Canadian Forces have, in fact, retained one not insignificant reinforcement commitment to the defence of Norway: the battalion group that would join the northern component of the Allied Mobile Force-Land (AMF-L). This battalion group is based on the 1st Battalion of the Royal Canadian Regiment (1RCR), located at CFB London, Ontario that, along with "slices" of other units, would be airlifted to

> *the Northern Flank. The AMF-L is a multinational, brigade-sized, quick reaction unit.*
>
> *Equipment being pre-positioned in the Tromso area of North Norway originally intended for the lead battalion of 5GBC in its CAST guise is still being kept there for emergency use by the Canadian AMF-L battalion group. It would thus not be inconceivable for the Canadian government to cancel entirely the plans for a Canadian division in Germany and restore the CAST commitment to its 1987 level.*
>
> *Nonetheless, the 1987 level is not good enough. Beyond the transit time to Norway of up to one month for 5GBC's equipment, enormous medical and logistic troubles attended the CAST commitment, as BRAVE LION demonstrated. Yet it cannot be stressed too strongly that these problems arose in large part from Canada's attempting to focus on the defence of two NATO areas, Germany and Norway, despite the limitations of the Canadian defence budget. As the government put it in its White Paper, "It has been obvious for some time that these widespread land and air force commitments in Europe represent a dilution of valuable combat resources, and cannot easily be supported or sustained from an ocean away in the event of hostilities. They force us to maintain widely separated lines of communication for which we have insufficient strategic transport."*

The professors are not saying that Norway is too hazardous. It is also true that forces left on the Central Front would also be vulnerable if inadequately prepared. The Norway location may in fact be less hazardous than the one in the Black Forest.

I would suggest, as the professors have done, that the White Paper should now be brought to its logical conclusion. It is possible to re-structure the land force commitment to Europe without leaving Europe altogether, without abandoning Norway, and without pretending to have resources that we have no intention of acquiring. We can be in Europe and be seen to be there without maintaining so large a permanent presence there.

Canada's forces in Europe have long been regarded as a political necessity. The Americans and Europeans have tended to gauge our commitment to the Alliance by the quantity and quality of the forces located in West Germany. By maintaining a token force there Canada has qualified for "a seat at the table" for alliance and arms control discussions.

Those forces in Europe should not be the sole measure of our commitment. Europeans tend to forget that Norway is at the eastern end of NATO's northern flank, which extends across Canada and includes Alaska. What Canada does to protect its own territory and the American strategic forces is also a contribution to western defence.

The force that we maintain in Europe is small and could be more efficiently employed elsewhere. The three squadrons of CF-18s based in Baden-Soellingen could be redeployed in Canada for use in a NORAD and drug interdiction role. The two rapid reactor squadrons, based in Canada but committed to a European role, could continue in that role, but their area of operation could be changed to Norway. In that way they can be used for Canadian peacetime tasks but still be available for use in the European theatre in an emergency.

The ground forces could be withdrawn to Canada and dedicated to a North Norway defence role. Plans to purchase a new main battle tank could be cancelled, as troops in North Norway would be more lightly equipped. Infantry could be supported by artillery and a smaller number of tanks in the more restricted valleys of that mountainous country. As they age and become fewer by attrition and cannibalization, the Leopard I tanks could continue to serve the needs of this force, pre-positioned there. Alternately, an arrangement might be made with the Norwegians for provision of tanks for Canadian use.

The defence of Canada is primarily a task for the militia. This should be given emphasis, and may require the re-opening of armoury and other training areas across Canada. In any event, the size of the militia should be allowed to grow in accordance with the intent of the White Paper.

In order to achieve these objects, we should:

Seek permission from NATO to re-define and reorganize our European commitment as follows:

Restore the CAST commitment to reinforce North Norway in time of crisis.

Continue participation in NATO exercises in Europe.

Reduce the permanent strength at CFB Lahr to that of a staging base, with resources to support units that arrive there for training exercises; or better still, strike a deal with the Germans to trade the Lahr base for a staging base in northern Germany for the rapid reinforcement of Norway, in exercises and in time of emergency.

Point out to the Europeans that Canada's presence in Europe will continue, although on a different basis and that, in any event, Canada's contribution to NATO consist of the defence of our own country (as theirs do) as well as what we do in Europe.

Make greater use of pre-positioned equipment, including the use of allied equipment, and ensure that Canadian troops participate, on a fly in basis where required, in appropriate exercises with NATO allies, either in Norway or on the Central Front.

Sell off unnecessary infrastructure at Lahr.

Increase airlift capacity, to improve the CAST capability.

Buy no more tanks at present.

Increase the size of the militia and assign it specific tasks, with emphasis on territorial defence and on peacekeeping.

Increase the forces designated for peacekeeping operations, ensuring that Reserves form a larger part of the mix, and including training in counter-terrorism.

> GOD AND THE SOLDIER WE BOTH ADORE IN TIME
> OF DANGER, BUT NOT MORE DANGER OVER,
> ALL IS REQUITED. GOD IS FORGOTTEN,
> THE SOLDIER SLIGHTED. – anon.

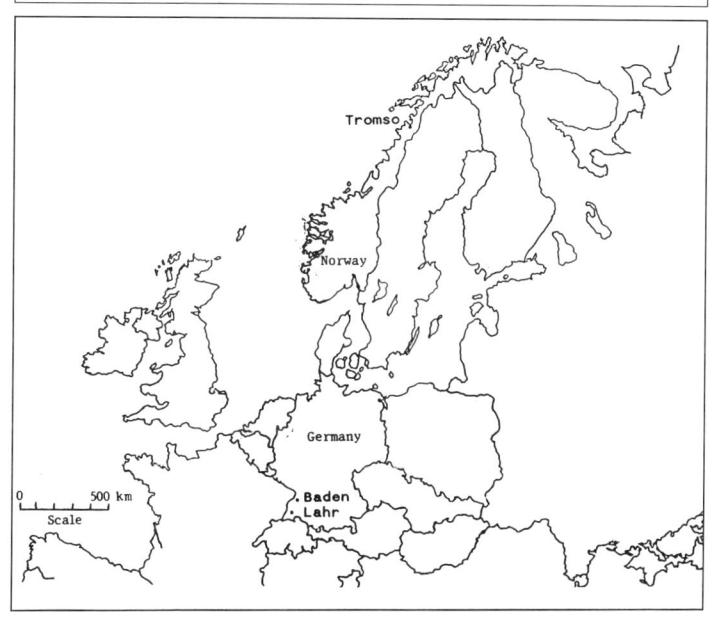

Western Europe, showing CF Bases at Lahr and Baden-Soellingen and CAST destination at Tromso, in North Norway.

3. Air Forces

Canada has a proud history of aviation. Our First World War aces, the pioneering bush pilots of the 1920s and 1930s, the large numbers who served with distinction in the Second World War in bombers, fighters, transport, coastal and ground operations - all brought this country honour and a reputation for professionalism. The Commonwealth Air Training Plan, the largest and most successful ever known, was an important factor in winning the Second World War and a remarkable feat of organization and production.

Like the sister branches of the service, the air force was reduced in size during peacetime. It began the 1990s with the following, according to information supplied by Chief of Defence Staff in November, 1990.

Table 8.

AIR FORCES

Personnel	
Regular	22,000
Primary Reserve	1,500
Major Operational Units	
Tactical Fighter Squadrons	8
Maritime Squadrons*	9 (1 Reserve)
Tactical Helicopter Squadrons**	8 (4 Reserve)
Medium Transport Helicopter Sqdns**	2
Transport and Rescue Squadrons	6 (2 Reserve)
Transport Squadrons	6
Radar Squadrons	6

Principal Equipment

	in Canada	in Europe
Tactical Fighters(CF-18)	71	42
Tactical Fighters(CF-5)	64	0
Maritime Aircraft*	62	0
Tactical Helicopters**	88	13
Transport Helicopters**	7	0
Tactical Transport Aircraft	39	1
Strategic Transport Aircraft	5	0
Search and Rescue Aircraft (SAR)	48	0
Training Aircraft	221	5

Bases in Canada: 18

* Squadrons under operational control of Maritime Command
**One squadron under command of 4CMBG, other squadrons under operational control of Mobile Command

In recent years the roles have been varied and demanding. At present they are:

- aerospace surveillance and defence of this country and North America,
- maritime surveillance and defence,
- territorial defence,
- NATO collective defence,
- peacekeeping.

The activities include: North Warning System enhancement, including fighter operations from austere and remote airfields, anti-drug operations, intercepts of Soviet bombers in Canadian air space, Search and Rescue, humanitarian actions, such as the airlifting of supplies to Ethiopia, Armenia and the Kurds, support to the United Nations operations in Sinai, Iran-Iraq, Namibia, Central America and Kuwait, support to Fisheries and Oceans, Customs and Immigration, the Atmospheric Environment Service and in aid of sovereignty, wildlife counts and oil pollution control.[22] In 1989 our air force flew 282,000 hours in fulfilling its assigned roles and missions.

To make all this happen, Air Command, the former Royal Canadian Air Force, is the largest element of the Canadian Forces. Its 30,000 men and women consist of some 22,000 Regular Force personnel, 6,868 civilians and fewer than 1,500 Air Reserves.[23] It has more than 700 aircraft, and until recently these were of 23 different types - a costly variety in terms of maintenance, operations, training and logistics.[24] By early 1991 the number of types had been reduced to 15.

As of 1 November 1989, the 1st Canadian Air Division consisted of the 3rd Fighter Wing, based at Lahr, and the 4th Fighter Wing at Baden-Soellingen with its three tactical fighter squadrons of CF-18s, two rapid reinforcement squadrons and an air maintenance squadron. There are six air groups based in Canada: Fighter group, with its CF-18 Squadrons at Cold Lake and Bagotville, Maritime Air Group which provided MARCOM with Aurora long range patrol craft, Sea King helicopters and(until recently) Tracker patrol aircraft, 10 Tactical Air Group which supplies helicopters in support of Mobile Command, 14 Training Group, the Air Reserve Group and Air Transport Group.

[22]Canada's Air Force, a presentation by LGen F.R. Sutherland at the 1990 Conference of Defence Associations (CDA) Institute Annual Seminar.
[23]Defence 89 p. 50.
[24]Canada's Air Force - Sutherland.

In Europe in 1987-88 1 Canadian Air Group, as it then was, participated in 20 NATO exercises, in addition to monthly fighter aircraft exercises. In the following year concentration was on air-to-air and air-to-ground roles and on "survival to operate". Crucial areas of training there that year were on individual skills such as nuclear, biological and chemical defence, explosive ordnance disposal and rapid runway repair; practising procedures for the transition from peace to war; and ensuring that each pilot flies at least 240 hours a year, which is the minimum required to maintain combat readiness. Since then the three squadrons have all been equipped with the multi-role CF-18 fighter aircraft. With the flip of a switch it can be changed from an air-to-air role to an air-to-ground role.[25]

The CF-18 was the most advanced piece of equipment in the Canadian arsenal and available for service in the Persian Gulf in 1990-1991, and it performed admirably there.

The same questions that arise in connection with the land forces also apply to Canada's air force in Europe. Should we remain in Europe? Should we shift the emphasis and the roles? And should we still be in the high-performance fighter aircraft business? In view of the government's deficit problems, it is tempting to conclude that no more CF-18s should be acquired, although in time the attrition of aircraft would then make it impossible to maintain our current commitments to NATO.

If the aircraft were to be brought home to Canada this act would bolster the country's home defence. But it would also tend to remove the air force further from NATO contacts and place it more firmly than ever in the role of junior partner to the US Air Force in the defence of North America. Our air force has for many years been more closely identified with American thinking and equipment than either the army or navy, and moving the squadrons home from Europe would strengthen the connection. Is that what Canada wants or needs?

On the other hand, we are reminded that the reason defence forces deteriorate in time of peace is because they have failed to find peacetime missions. When the public senses an absence of danger, the mood is not right for battleships, aircraft carriers or large numbers of tanks or fighter planes. But Canada's air force has missions and roles at present that are highly suited to this time of peace and that can also keep the necessary skills alive for the

[25] 1989-90 National Defence Estimates, Part III p. 55.

time they might, regrettably, be needed again for hostilities. We should be building on those missions, those roles, those skills.

The Air Transport Group and Maritime Air Group perform tasks that are essential for surveillance and the defence of Canada and that project this country's will to assist other countries in humanitarian and peacekeeping activities. Maritime Air Group needs more Aurora LRPAs than it now has, to perform a credible job of sovereignty patrols; the Trackers need to be replaced if we are to be serious about inshore fisheries patrols and surveillance for drug interdiction and illegal immigration. Search and Rescue is a valid and honourable role, as is peacekeeping airlifting. If Canada were to maintain fewer personnel and less equipment in bases in Europe, it would be possible to divert the funds saved toward the purchase of additional aircraft for airlifting troops on exercise or for rapid reinforcement. It might even be possible to boost the Canadian aircraft industry in the process.

Equipment for Canada's air force is expensive: the CF-18, for example costs more than $35 million per copy. The Long Range Patrol Aurora has been mentioned before and is also costly. But the workhorse planes of Air Transport Group, such as the C-130 Hercules or the Boeing 707 seldom figure in controversies over

Boeing 707 tanker plan refuelling two CF-18s.

costs. Nonetheless these are the aircraft that carry Canadian forces, fly relief missions and supply the transport for major movements for Mobile Command. These are the aircraft that come closest to meeting the test of a convincing, viable peacetime role, and there are never enough of them.

> BUT WHAT WOULD BE THE SECURITY OF THE GOOD, IF THE BAD COULD, AT PLEASURE, INVADE THEM FROM THE SKY?
> Samuel Johnson, 1759

NORAD

No comment on Canada's air forces can be complete without a reference to NORAD, the North American Aerospace Defence Agreement. NORAD is the most visible of Canadian-American defence arrangements.

The story of defence co-operation between our two countries begins, for most purposes, in 1938 when Prime Minister Mackenzie King and President Roosevelt signed the Ogdensburg Declaration enshrining the vows that the United States would "not stand idly by" if Canada were to be attacked and that, on Canada's part, "should the occasion ever arise, enemy forces should not be able to pursue their own way by either land, sea or air to the United States across Canadian territory".[26] The Permanent Joint Board on Defence (PJBD) was formed that year.

Other milestone events included the building of the Alaskan Highway across Canadian territory in 1942-43, the 1949 creation, under NATO, of the Canada-US Regional Planning Group (CUSRPG) and the 1951 agreement to build the warning line known as the Continental Air Defence Integrated North (CADIN) - Pinetree line of radar stations. The military plans prepared for the Military Co-operation Committee of PJBD and those of CUSRPG are almost identical, and little wonder. NATO's description of the latter is terse: "The Canada-US Regional Planning Group, which covers the North American area, develops and recommends to the Military Committee (of NATO) plans for the defence of the

---

[26]Report of the Commons Standing Committee on External Affairs National Defence - NORAD 1986, February, 1986.

Canada-United States region. It meets alternately in one of these two countries".[27] To my mind it is perfectly appropriate to think of NORAD as the military co-operation arrangement for the air defence of the North American region of NATO.

NORAD is a fact of Canadian defence life. It is sometimes said that in NORAD we Canadians are a junior partner to the US. But the alternative is not to be a partner. As partners we participate in plans and activities, and we have an avenue for voicing our views on the fate of the continent we share. NORAD is a fact of life. It should remain so and continue to enjoy the enthusiastic participation of level-headed Canadians. Apparently the government agrees, as it is renewing the agreement for a further five years - until 1996.

National Air Surveillance

In this country, facilities for the surveillance and control of Canadian airspace cover only those areas of specific interest to NORAD, and are provided and managed by the Air Force. This is consistent with the general direction given by the Aeronautics Act, which makes National Defence(DND) and Ministry of Transport(MOT) both responsible for the surveillance and control of the airspace above all Canadian territory. The degree to which these tasks are carried out by the two agencies is a measure of the sovereignty Canada has over its own airspace.

The Act leaves the details of the sharing of that responsibility to bureaucrats in the two federal agencies, who have jealously guarded the "turfs" of their separate organizations. As a result, Canada does not have an overall Air Surveillance Plan which would form the basis for intelligent sharing of responsibility in the national interest. There has been a duplication of costly air surveillance and control facilities, leaving no funds available for coverage over other Canadian airspace, such as over the Canadian Archipelago, and until recently along the coast of Labrador. The range of the new North Warning System extends to only 72 degrees N, which is south of the North West Passage through Melville Sound. The Canadian decision to limit its air surveillance of the Archipelago to meet only the needs of NORAD raises the question: How can Canada claim political sovereignty over land for which it has already surrendered military sovereignty?[28]

[27]NATO Facts and Figures, 1984.
[28]Minutes of Proceedings and Evidence of the Standing Committee on External Affairs and National Defence, 17 September, 1985, pp. 28.8 - 28.9.

The Aurora long range patrol aircraft must cover vast expanses of Canada's north and three coasts.

Canada has only "Paper Sovereignty" over its airspace north of 72 degrees N, an area the size of the British Isles. This area is overflown by 6,000 aircraft annually that file flight plans and report their progress along those High Arctic routes. Canada has no means of knowing whether any others use this airspace. Actual sovereignty would be ours if we knew what other aircraft fly in that area. That would enable effective control in that airspace claimed by Canada.

Proposals to provide the necessary surveillance over the Archipelago have been dismissed as being too costly. Concurrently, MOT's Airspace System Plan to the year 2000 is being implemented, and DND is upgrading the Distant Early Warning(DEW) line with the North Warning System, and extending it to cover Labrador to meet NORAD's needs to protect the nuclear deterrent. These separate plans are still to be completed.

The sums being spent by the two agencies are substantial: about $5 Billion by MOT by the year 2000, with $2 Billion in the first phase by 1992, and about $1 Billion by DND by 1991.

Other countries have successfully avoided the high cost of duplicating civil and military systems for surveillance and control of their airspaces. In the United States and France, for example, and soon in Australia, radar installations, communications systems, and control centers are combined. Those examples and commonsense indicate that a combined surveillance and control system in Canada would save considerable federal funds, and not increase the national deficit.

It should be mentioned at this point that, having closed most of the radar stations in its Pinetree Line, Canada is now largely dependent on that same US Joint (civil and military) Services System(JSS) for its air defence. In the High Arctic, Canada depends on intermittent surveillance by American AWACS(Airborne Warning and Control System) flights.

The root of the problem I have been discussing is Canada's Aeronautics Act. It makes MOT and DND both responsible for the surveillance of Canada's airspace. MOT is responsible for the control of cooperative aircraft, and DND for the detection and control of uncooperative aircraft, exercised jointly with the United States through NORAD. One cannot assume that by satisfying NORAD's needs the sovereign interests of Canada are also satisfied.

The intent of the Act should be stated more clearly to prevent federal agencies from continuing to use it as their authority to duplicate identical or quite similar radars, communications, and control facilities. The funds spent on two separate systems have had a negative impact on the provision of coverage within the perimeter of all the airspace claimed by Canada.

A National Air Surveillance and Control Plan for Canada is necessary to prevent the duplications and shortfalls already mentioned. The plan would enable Canada to negotiate with the US as one entity rather than through two separate agencies. A National Plan would accelerate the establishment of Canadian sovereignty over the true perimeter of Canadian airspace, including our Arctic regions.

The urgent need for such a National Plan was highlighted by a working group of the Canadian Institute of International Affairs in 1988.[29] Canada cannot "declare it has full jurisdiction and control of its own airspace unless it has the capability for identifying and controlling all air traffic in it."

There is a need for a more responsible attitude by government agencies in planning and providing facilities for the surveillance

---

[29]*The North and Canada's International Relations. The Report of a Working Group of the National Capital Branch, Canadian Institute of International Affairs.* Published in 1988 by the Canadian Artic Resources Committee, Ottawa.

and control of the airspace of Canada.[30] Therefore the government should:

> Prepare a National Air Surveillance and Control Plan for Canada, with the mandate that it include all Canadian airspace.
>
> Clarify the responsibilities of federal agencies, including the Ministry of Transport and the Department of National Defence, for the surveillance and control of all Canadian airspace, by amendments to the Aeronautics Act of Canada.

In summary, therefore, in order to keep Canada's air force alive and healthy, to perform the aerospace defence of Canada and North America, to support other important military missions, to continue our role as good international citizens, to provide aid to the civil power, assist the civil authorities, to contribute to the maintenance of stability around the world, to enhance Canada's image through peacekeeping and humanitarian undertakings, to support national events, to support the national strategy for the war on drug trafficking and to do all this while keeping the expenses under control, it is suggested that the following measures would be appropriate:

> Acquire no additional CF-18s in the near future.
>
> Deploy the CF-18s in the defence of Canada and North America, and for rapid reinforcement of NATO forces in North Norway.
>
> Continue participation in NATO exercises, and in NORAD exercises.
>
> Replace and augment the transport aircaraft, both fixed wing and helicopters, to enable an expanded capability to airlift troops for exercises and to support NATO roles, peacekeeping and other UN and humanitarian activities.
>
> Augment the LRPA capability with additional Aurora or comparable aircraft for surveillance of Canada's coastal and open waters.

---

[30] *The Arctic Group: Choices for Peace and Prosperity*, Proceedings of a Public Inquiry, T.R. Berger, Soviet Ambassador A. Rodionov, D. Roche and 21 other speakers, the True North Strong and Free Inquiry Society, 1989.

Replace the Tracker patrol planes for inshore surveillance, for fisheries patrol, drug interdiction and detection of illegal immigration with aircraft suitably fitted for the role.

Continue the process, already begun by Air Command, of rationalizing the fleet by reducing the variety of aircraft.

Reinstate the program to modify all seven Challengers for electronic warfare training.

Improve the efficiency and cost-effectiveness of Canadian air surveillance services by clarifying both the separate and shared responsibilities of the Ministry of Transport and National Defence, by amending the Aeronautical Act of Canada. Canada should contribute joint surveillance information to NORAD.

Extend the ground-based surveillance of Canada's Northern and Arctic regions to cover all territory claimed by Canada

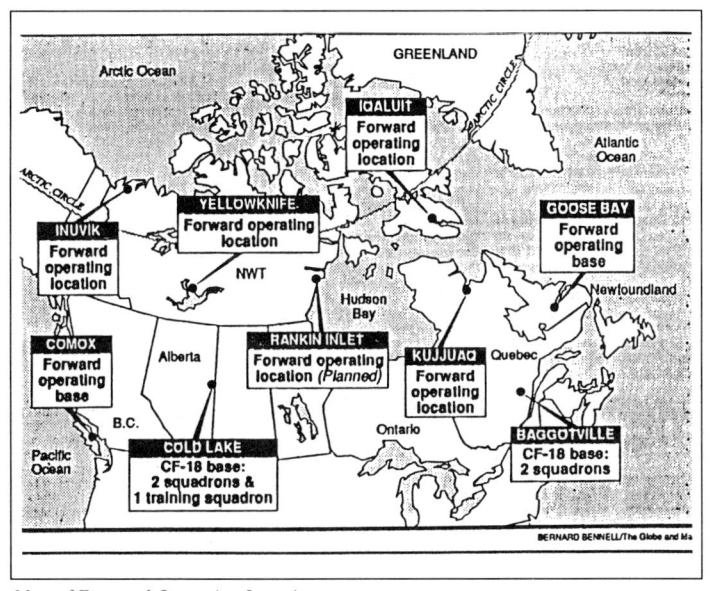

Map of Forward Operating Locations

4. Reserves and Cadets

The first thing worth mentioning about Canada's Reserve forces is that they have been so few in number. In most of the world's developed countries the Regular Force is modest in size while the Reserves provide the numbers that permit mobilization in time of need. The International Institute for Strategic Studies comments that Canada's "primary reserve units are well below war strength and are normally composed of Unit HQ, HQ elements and one sub-unit".

Table 9

Selected Comparisons of Military Manpower in 1988 (000s)

	Regular Force	Reserve Force
Belgium	88.3	145.0
Denmark	29.3	74.7
France	456.9	356.0
FRG	488.0	850.0
Greece	214.0	404.0
Italy	386.0	769.0
Netherlands	102.2	170.3
Norway	35.8	200.0
Portugal	73.9	190.0
Spain	309.5	1,030.0
Turkey	635.3	951.0
United Kingdom	316.7	319.8
Canada	84.6	21.3
Sweden	67.0	609.0
Switzerland	3.5	601.5
USA	2,163.2	1,675.8
USSR	5,096.0	6,214.0
Non-Soviet WP	1,211.3	2,060.5

Source: The Military Balance 1989-90. IISS.

The ratio of Reserves to Regulars is particularly high in the neutral countries of Sweden and Switzerland. In the latter, especially, the tradition of the citizen soldier is very strong.

But in Canada, as well, the role of the Reservist deserves more attention than we have seen in recent years. The term Reservist includes, of course, the Militia in the Land Forces. In a democracy every able-bodied citizen has the right to serve and the duty to serve in the protection of his or her country. The Reservist has two lives, the main one is in the civilian world; the secondary one is with the armed forces of his or her country. Participation by part-time soldiers, sailors and air force personnel, always in contact with the community, is one of the best ways of assuring that militaristic influences will never take hold in a free country. Thus Winston Churchill was once led to remark that every Reservist is twice a citizen.

Reserves are cheaper to maintain, and they thus provide an economical way to field the numbers that may be required for emergencies, military or civil. And in Canada we remember that our wars have been fought largely by Reserves - an idea that is no longer considered out-of-date.

When addressing the House of Commons Standing Committee on National Defence, the Director General of Reserves and Cadets had this to say:

> *"The members of the militia are a unique segment of our society. They are citizen soldiers who balance their obligations to their family, business and community with those of their nation. They are proud of the units with which they serve and the service which those units have performed in the past in the defence of Canada."* [31]

How effective are Reserves in meeting Canada's defence needs? Ask different people and you will get different answers. Some Regular force personnel have occasionally referred to Reserves in unflattering terms, such as clumsy, inept or incompetent. This no doubt reflects the mythology of professionalism and an inability to concede that a part-time soldier could ever measure up to the Regular. Ask a Reservist the same question, and the answer may well be "the backbone of the Forces".

The criticism, if intended as a generality, is unfair and inaccu-

---

[31] Brigadier General Larry Gollner, DGRC, Dec 1, 1987.

rate. It fails to recognize the potential for very high competence that lies within each part-time member of the Forces. Given a chance at the same training, Reserves have often outshone their fulltime colleagues. When the National Defence Speakers Bureau included Reserve officers in its program of seminars to develop speakers, the performance of the Reserves was extremely impressive. Usually, all the Reserve needs is more time to learn the knowledge and skills and to put them into practice.

The purpose of Reserves in Canada, after all, has never been to produce a parallel, fully-trained branch of the service. The purpose of Reserves has been to develop a partly-trained cadre that could, when needed, be called upon to upgrade its skills and put them at the service of the country. Thus, the fine programs for officer development at universities that flourished during the 1950s and 1960s, such as the University Naval Training Division (UNTD), the Canadian Officer Training Corps (COTC), and the University Reserve Training Flight (URTF) all served their purpose admirably. Some of their graduates became members of the Regular forces; some continued active in the Reserves after commissioning; and others, even if they never wore the uniform again, still brought to their communities, professions and families a better appreciation of defence problems.

Navy, Army, and Air Cadet Leagues at some 25 locations in Canada provide training in leadership, physical fitness, and in activities such as flying, sailing, communications, parachuting, and a number of trades during the school year and at DND summer camps, which 20,000 Cadets attended in 1989. Cadet programs at all levels, especially for those at the secondary school age, are not intended as recruiting bases. Their purpose is to develop good qualities of citizenship, and a sense of national unity and pride in young Canadians. Today we need more of these programs.

With the defence cutbacks that began in the 1960s the numbers of the Reserves fell drastically as Reserve training establishments and armouries were phased out. One of the results was the loss of contact with the community, as mentioned in an earlier chapter. Another is the apparent tendency to spend large amounts on fewer members of the group. Is it really cost-effective to send groups of Naval Reservists from say, Moose Jaw or Winnipeg, travelling by commercial air for a weekend of sea training out of Esquimalt?

Surely it would be better, and more cost-effective, to arrange periods of a week or more, provided the leave can be arranged. The leave, of course, is a problem.

The Government of Canada has provision for its own employees to take "military leave" without prejudice to other annual vacation time. Unfortunately, this is still rare among civilian employers. The result is that the Reserves depend heavily on students, teachers and others with long periods of annual leave, and on the unemployed. Although there are exceptions to this, there obviously exists a need to make Reserve service more attainable for a better cross-section of the community. At present there are difficulties facing different segments of the population. High school students, for example, have time for training but they tend to leave the Reserves when they leave school. Nineteen-year olds in the work force, on the other hand, tend to stay with it, but have a greater problem in finding the time for periods of concentrated training. So there are not enough of them.[32]

DND has identified the need to persuade the private sector to make their employees available to perform military service, and the National Employment Support Committee(NESC) has made senior executives more aware of the military demands in effort and time put on Reservists. More effort is needed, combined, perhaps, with the right incentives.

The pay of a Reservist has also been a problem - the amount and the promptness with which it arrives. Evidence presented to SCOND in 1987 revealed that Class A and B Reservists, those who turn out for weekly parades and weekend training receive much less pay than those who serve on extended Class C duty or as Regulars.

"The cost for pay comparability at current personnel levels has been estimated at about $50 million annually in 1987-1988 dollars".[33] The pay differential has begun to close. Wages of Reservists have been increased an average of 14 per cent since publication of the White Paper in 1987, to bring them closer to the wages of the Regular Force. Pension, death and disability benefits, comparable to those of the Regular Force, are being developed in a total health care package for the Reserves.

Young Canadians appear ready to join the Reserves, whether for patriotism, for adventure or to learn useful skills; but they cannot be induced to remain unless these problems are addressed.

[32]Col. Brian MacDonald, addressing Senate Special Committee on National Defence, May 17, 1988.
[33]SCOND "The Reserves", June 1988.

The role of the Reserves has been the subject of much discussion within defence circles. The traditional role, as the SCOND report of 1988 observed, has been "to form the basis for augmenting the Regular Forces in wartime". But there are other views, those that argue for specialized roles for Reserves. A case can be made for both views, the augmentation role or the specialized function. One of the arguments for a specialized function is that modern technology is in some cases beyond the mastery of Reserves in the time available to them. At sea, for example, a modern destroyer or frigate requires a majority of personnel to have a technologist's training or better, and the air and land forces have sophisticated weapons and operational equipment, unlike the days of yore when, with a three-or-four week course in seamanship a young seaman could start to earn his keep on the upper deck, or an infantry soldier could be made partly ready for battle in the same space of time.

In the 1950s, the Royal Canadian Navy identified the role of Naval Control of Shipping as one that Reserves could readily assume. It is a role needed only in time of emergency and is one that can be practised and trained one or two evenings per week. More recently, the acquisition of the first two mine countermeasure vessels, HMCS Moresby and HMCS Anticosti, marked the beginning of another new role for naval Reservists. Minesweeping and minehunting make fewer technological demands than the blue-water warships. Convoy commodores may also be trained from within the ranks of Reservists as well as retired Regular Force members.

However, there will always be many to argue the case for the augmentation role. To some it is a matter of wanting the Reservist to experience the same work and adventure as the Regular. For many, the appeal of joining the Reserves was and is the opportunity to serve in the same ships, tanks, trucks, planes,etc., in other words, to be in "the real thing". For others there was the concern that a specialized role could end up being the same old story of having to endure outdated equipment and other forms of discrimination. The "Total Force Concept" announced in the White Paper was intended to ensure that all Regular and Reserve members receive equipment when it is new, so that all will be familiar with it and so that Reservists may join the Regulars at the drop of a hat.

It seems to me that both approaches are valid and compatible, and that the concept of the Total Force ought to embrace both.

Maritime forces had an authorized paid ceiling of 4,212 Reserves for 1990-91. The Naval Reserve is being trained for the command and operation of Maritime Coast Defences and the Naval Control of Shipping, integrating Regular and Reserve Forces. Members of the Naval Reserve have participated in Mine Countermeasures training with the Royal Naval Reserve and The French Navy. They now man, year-round, the two vessels for minesweeping (now known as Maritime Coastal Defence Vessels) pending the delivery of newer ships. With two new Naval Reserve Divisions the number of Naval Reserves is expected to grow steadily. The Total Force Concept ought also to allow Reservists to transfer to Regular Force positions, whether in peace or war.

The authorized paid ceiling for 1990-91 for the Militia in the Land Forces was 20,105. Their role is the Territorial Defence of Canada, and as well they perform ceremonial functions. They have trained in exercises in Canada and Europe, and a few have served in peacekeeping operations in Cyprus, the Golan Heights and Namibia where, in all these activities, their presence contributed to the success of the Regular Forces. The militia perform local search and rescue operations, as well as base defence and augmentation of Regular units in which they are being integrated in the Total Force concept. But could the militia not do better than that? The Reserves of many other countries serve as front line troops in Europe, mostly as conscripts fulfilling their country's defence commitments.

Picture this scenario: militia units from say, Ontario or Manitoba exercising regularly at Cambridge Bay or other similar locations in the Arctic, their presence there enhancing Canadian sovereignty, as do the flights required to get them there. With such experience, they can then also be used in Northern Norway with greater confidence than if all their domestic experience had been in the "south". At the same time, Air Transport Group would develop more expertise in cold weather hauling.

The Canadian Rangers are 1700 Reservists native to Canada's Northern regions (Arctic and Labrador) who assist in the selection of suitable areas for exercises by Reserve and Regular Forces from the "south". They act as guides and survival instructors, and assist in search and rescue operations. They are a mobile force who reconnoitre large portions of the Northern Region, which has its headquarters at Yellowknife and covers more than one third of Canada's land and fresh water. A Pacific Coast Region has been proposed, to provide Ranger patrols for all of British Columbia.

The commitment for new armouries and naval training centers that resulted in ten new facilities since 1981 in home communities has been increased, and it is a pleasure to note that more are to be constructed each year. Only when there are more young Canadians visible in their communities, proudly wearing their uniforms, will the link be restored between the Canadian Forces and the people.

The Air Reserve Group is very small, with an authorized paid ceiling in 1990-91 of only 1472. As the SCOND observed in 1988, its main problems were those of lack of personnel, inadequate training facilities and aged equipment (they were then still using the DC-3, but have since acquired the DASH 8). The Air Reserve Group now has permanent units in Europe, has participated in exercises of Air Command and has brought its training standards closer to those of the Regular Force.

The Communications Reserve, within an authorized paid ceiling of 1,712 (1990-1991 figures)[34] reinforces communications units of the Regular force, with which it has been integrated since 1970, and trains signallers in the Reserve units of the other Commands. Its operational importance was recognized in 1989 by the creation of a Reserves Branch to develop No.79 Strategic Communications Regiment. This unit will be capable of providing permanent communications services for the UN and NATO, and rapid communications for defence in Canada and aid to the civil power in domestic emergencies. Members of the Communications Reserve have served in peacekeeping operations, where Canadian communications are respected and in demand.[35]

---

[34]Defence 90

[35]For more information on Reserves and Cadets, see *Defence 90*, published by National Defence.

The White Paper goal of 90,000 Reserves should be restored, beginning with a drive to raise the militia to 30,000 from its present 20,105.

Reserve facilities should be re-opened where necessary, and the construction program should be expanded, especially in those new-growth parts of Canada that have none, in order to allow a military presence and to establish links with those communities. Facilities need not be lavish.

The University officer training programs (UNTD, COTC, URTF) should be restored and supported.

DND should ensure that Reserve activities and Cadet activities are well publicized.

Cadet training programs, with their proven capability to develop good qualities of citizenship, a sense of national unity and personal pride among young Canadians, should be expanded.

The program to arrange for Reserve personnel employed in the private sector to be allowed leave for military training purposes should be vigorously pursued.

The process of making the pay and basic pension, disability and death benefits of the Reserves comparable to those of the Regular Force should be quickly concluded.

The Total Force Concept ought to be continued, while allowing for specialized roles such as Naval Control of Shipping and Mine Countermeasures for Naval Reserves and Land force units concentrating on northern training, anti-terrorism and peacekeeping.

V LEADERSHIP ISSUES

A. Policy

"The relevance of the experience (of 1914 to 1918 was) that the military institutions we will need in an emergency will have to be prepared in peacetime, for it is only then that we have the time to develop experienced leaders, both in the ranks and as officers. Goodness knows how limited and often irrelevant peacetime training and experience may be, but it will always prove more valuable than no experience at all".[1]

The 1987 White Paper on defence was the first such document to appear in 16 years. But in 1988 it was necessary to issue an Update document in the light of events in Europe that were already beginning to change. Then the Budget of April 1989 superseded all, with its announced cancellation of major defence programs and plans. Since then, defence decisions have been taken on an ad hoc basis and the military have been left to cope with the impact of international and domestic events with no political guidance. With the Cabinet pre-occupied with other matters it appeared to observers in 1990 as though the government had no policy at all and that the government was not concerned enough to produce one.

When the Minister of National Defence ordered DND speakers, military and civilian, to go forth and explain the government's policies, in 1986, 1987 and 1988, where were the Cabinet Ministers? Why did not the Prime Minister and the Secretary of State for External Affairs get out in public, to explain and sell the White Paper? When the axe fell the Minister of National Defence was left out on a limb, and the DND staff, military and civilian, felt abandoned and betrayed.

Since then the politicians have been reluctant to write a new defence policy because of the criticism directed at the 1987 White Paper, especially in the light of succeeding events at the Arms Control negotiating tables in Geneva and Vienna and in the weakening Warsaw Pact. But a quickly changing international environment should not be an excuse for doing nothing. Instead, a defence policy should be fluid enough to respond to changes, both

---

[1] Dr. Desmond Morton, Principal of Erindale College, University of Toronto, testifying to Special Committee of the Senate on National Defence, April 26, 1988.

in Canadian interests and in the international environment, while providing the military with the direction it needs to plan its long-term programs.

If it shows nothing else, the experience of hastily assembling naval and air forces for service in the Persian Gulf in 1990 should have shown that what we must always expect is the unexpected. Defence in Canada needs a long-term commitment, preferably in the form of a new defence policy and an annual update.

B. Social Issues

Long ago, before the days of nuclear power and laminated armour, when ships had sails and cavalry meant horses, there was little talk of human rights or making the military reflect the modes of society. Press gangs found recruits in taverns, and if a man could walk, he was fit enough to go to sea. And they were men; the only women in the army were the camp followers. As for what is today called sexual preference, the navy called it buggery and it was a heinous sin, cause for instant dismissal if the offence took place in or near the ships.

Today the military is expected to function in a different social and political environment, one in which the rules appear to be changing rapidly and in which the perceived needs of defence do not seem to be a consideration. Policies the government wishes to promote (such as bilingualism) or pressures the government chooses not to resist (religious and cultural identity) are thrust upon the defence community, seemingly without regard for their impact.

Bilingualism became policy in the late 1960s and was extended to the Canadian Forces in various ways, including FLUs, or French language units. Thus, for example, one of the navy's ships was designated a FLU - a ship in which francophones could operate entirely within their own language. Later, a second ship was made a FLU but, when HMCS Algonquin was visited in 1985 by naval reunion tours the ship was FLU in name only. Although signs were in French, spoken commands were being given and understood in English, and most of the crew were English-speaking. The navy had been unable to find enough francophones who were willing to serve in an all-French ship. They had gone a ship too far.

Homosexuality was simply not tolerated. Until recently it was cause for denial of a security clearance, in the forces as in civilian departments of government. In the forces it was considered disruptive, a handicap that a fighting unit could do without. But today one encounters members of the forces who wish to come out of the closet and stay out, and they are contesting dismissal and transfer directives.

Women, who in the 1940s and 1950s were employed primarily in clerical and other support roles, are now eligible - as of February, 1989 - for all roles in the forces, including combat, with the exception of submarine service. Among the 16 countries of NATO Canada has the second highest participation of women at all levels, as Table 10 indicates. Women may today be found at sea, in fighter planes, graduating from military colleges and in many other non-traditional roles. But, as reported to the Senate, some Army combat trials of mixed gender units were unsuccessful.[2]

Table 10.   Statistics on women in the armed forces

	Belgium	Canada	Denmark	France	Germany	Greece	Italy	Luxembourg	Netherlands	Norway	Portugal	Spain[1]	Turkey	UK	US
1.a. Total number of men and women in the Armed Forces	92,845	85,692	29,646	556,500	487,000	165,084	..	734	106,204	41,476	75,032		749,003	312,744	2,123,669
1.b. Total number of men and women in the Armed Services less Compulsory Conscripts	57,184	same as a.	21,691	303,710	261,920	N/A	..	same as a.	57,222	13,142	31,120		85,588	same as a.	same as a.
2. Total number of women in the Armed Forces	3,379	8,328	1,032	20,470	208	2,746	..	16	2,279	768	9		8,014	16,022	221,649
3. Percentage of women in the armed Forces:															
a. Percentage of total number (1.a)	3.6 %	9.7 %	3.48 %	3.7 %	0.043 %	1.67 %	..	2.18 %	2.1 %	1.8 %	0.012 %		1.07 %	5.1 %	10.43 %
b. Percentage of total less Compulsory Conscripts (1.b.)	5.9 %	same as a.	4.7 %	6.7 %	0.08 %	N/A	..	2.18 %	3.98 %	5.8 %	0.03 %		9.36 %	same as a.	same as a.
4. Total number of women officers	132	1,647	61	1,015	208	424	..	N/A	283	429	9 (Air Force)		1,565	2,405	32,651
5. Percentage of total officers	N/A	9.2 %	1.2 %	2.6 %	0.5 %	N/A	..	N/A	2.4 %	3.3 %	0.082 %		N/A	N/A	10.7 %
6. Total number enlisted women	3,247	6,681	971	17,385	N/A	2,340	..	16	1,996	318	..		6,449	13,671	188,998
7. Percentage of total enlisted	N/A	9.8 %	6.2 %	6.6 %	N/A	N/A	..	2.9 %	4.1 %	1.1 %	..		N/A	N/A	10.4 %

.. Figures not yet available   N/A: Not available   31 December 1988

Source: NATO Review October, 1989

[2]LGen J. de Chastelain, addressing the Senate Committee on National Defence, 26 April, 1988. p. 33 of Proceedings.

The Assistant Deputy Minister (Personnel) reported :

> *"The experience we have had so far in the Army is not so much the Army rejecting the idea of combat units integrated as it is women rejecting the idea of coming into combat units. The numbers we have had volunteer in the NCM, (non-commissioned member) capacity, is minimal, and in some cases zero. So we cannot conduct a trial, for example, in field engineer units because we have no applicants to go into field engineer units. We feel that we will have enough to go into the infantry, artillery and signals, but we will be trialing (fewer) units than we had hoped to trial initially".*[3]

In some cases the problem of equal rights reflects the other side of the coin: women in the forces who are upset at the prospect of being forced to serve in combat roles for which they had not originally undertaken a commitment.

Much of the debate concerning the role of women in the CF has created, without doubt, an added burden on CF time and emotional energy, often with little payoff in return. However, there is one respect in which the open door to women is welcomed by Canadian Forces. The demographics of Canada, like those of many other countries today, reveal a shrinking base for recruiting in the traditional age groups. Admitting women to virtually all military trades has the effect of doubling the recruiting base.

In his appearance before the House Committee in 1989 the Minister recognized that responding to these new social pressures can lead to additional problems for which there are no easy solutions. He recognized that, to require all women in the forces to be eligible for transfer to combat units or other duties could be a breach of faith with those women, a change of the rules under which they signed on. However, he also recognized that, to allow those women exemptions from the new policies would constitute another form of discrimination, saying:

> *"What we have now is a change in the tasks because of a ruling (of the Human Rights Commission). If one were to allow female members of the Canadian Forces to choose to continue in their present task, occupation or environ-*

---
[3] Ibid. p. 33 of Proceedings

ment, then the male members of the Canadian Forces would wish the same opportunity and indeed should be given the same opportunity when we are addressing equality within Canada's forces." [4]

Access to Information rights has imposed a burden on the military, who regard some of what they do as rightly secret, unfit for all but the eyes of those who have a need to know.

Parenting leave is available to men as well as women.

Cultural and religious considerations have impinged upon the military of Canada. True, there is no more compulsory church parade on Sundays. But the drive to recognize and bestow dignity on minority groups has found its way into the military environment, and sometimes it comes into conflict with the wishes of the military or of good sense.

To one who wears a uniform, its distinctiveness is in the uniform itself. The word "uniform" means all alike, and all members of the group wear the same clothing, other than badges of rank or of achievement. In recent years, however, one cultural and religious group, the Sikh community, has insisted (successfully) that individual Sikhs be allowed to affect their personal and cultural style of dress while in uniform and while members of units that wear the standard uniform.

We have witnessed, in recent years, situations in which militia units attempted to re-create as historical pageant the events of other centuries, but at the same time those militia units have come under pressure to include members of visible minorities in the uniformed pageants, even at the cost of historical inaccuracy. This pressure, whether it comes from governments or from citizen pressure groups, is unfortunate. If the objective of such pressures is merely to satisfy current political fashions it should be unacceptable to honest persons, suggesting as it does the Stalinist insistence that only those works of art that reflected the politically correct views of the day could be put on display. It renders the exercise pointless at best, a lie at worst. It also shows that many of the persons exerting these pressures view the military as only a source of employment, having no other value.

A well-known and respected Jewish organization, the B'nai B'rith League for Human Rights, has challenged in the Federal Court the DND policy regarding participation in Middle East mili-

[a] The Hon. Bill McKnight, addressing the House of Commons Standing Committee on National Defence, June 1, 1989.

tary operations. The League argues that Jewish and Arab members of Canadian Forces should not be excluded from peacekeeping teams, even if the military believe that the avoidable presence of Jews in Palestinian areas or Moslems in Jewish areas could be provocative to the local populace. The military can be excused if they feel that their problems are being ignored in favour of some other agenda. The military are themselves famous for an excellent piece of wisdom: the commander on the spot is the best judge of the local situation.

I suspect that if the B'nai B'rith were to think this one through they might feel differently. For example, would they even think of hiring a Moslem fund-raiser to approach Jewish donors - even if the Moslem were a very good fund-raiser? Would that be discrimination? Or would it just be common sense?

The lives of the military are not like those in other occupations. When they join the forces they commit themselves to work a set routine of hours but, if required, to work 24 hours a day, seven days a week until the crisis is over, and all without overtime pay. Leave provisions are good, and so is the pay; but the life of the professional soldier cannot ever be like that of anyone else in a democracy. No one else is routinely required to follow orders without question, even when there is dangerous work to be done, even to risking one's own life. The rules for promotion are the Forces' own. They sign up to go anywhere, to do what they are told, to find pride in their noble profession of arms. And they have always known that they are a breed apart. But now the civilian world is intruding into theirs, and the result is not always happy.

It would be fair to observe that the changes are not all coming from outside. The military mess - formerly the centre of social life, the place where the history and the values of the regiment were learned, the true home of the soldier - is not what it once was. On an average evening, many messes are now largely deserted. Soldiers, sailors and fliers are now more likely to leave the base or ship when the working day is done. Instead of meeting in the mess to discuss the events of the day, they tend to go to their homes, join their families, put on their civvy clothes and light the barbeque, do what their neighbours are doing, and go back next day to what is their job. The military career is not, for many of its members, the be-all and end-all of their existence, although it might have been for their grandfathers.

The point of this discussion is to say that, while members of Canada's Armed Forces are not today so insulated from the currents of civilian society as their predecessors of even a half century ago, they nonetheless find those pressures uncomfortable to accommodate. They sometimes feel that changes are being thrust upon them that have no justification in the logic of their operational requirements, that the pressures are political in nature and have been put forward to satisfy the agenda needs of others. And they sometimes wonder if anyone in political authority cares enough or has the courage enough to protect them from unreasonable pressures that may impair their ability to perform the job for which they were hired.

Without intending to debate further the merits of any of the issues raised - or others not yet raised - I suggest that thegovernment should recognize that the situation of the Canadian Forces is not like that of any other organization, that the duties and responsibilities of members of the forces are not like those of any other Canadian, and that government should assume more responsibility for protecting Canada's Armed Forces from politically motivated pressures that could risk impairing its ability to perform its job.

C. Budget

Canadian defence budgets are not what they once were. In time of emergency they have consumed 50% or more of federal spending. In time of peace, in the early 1960s, defence accounted for close to 25% of the pot. Those were the days when we had more than 60 ships in the navy, there were active militia units across Canada, and the air force had four Wings in Europe as well as eight Reserve Wings in Canada. But then the cutbacks of the 1960s and 1970s resulted in a gradual shrinkage. The defence budget slipped to near 10% of the total, and the fighting ships, tanks, aircraft and personnel slid to the levels we know today.

So when the question of a peace dividend arises, it is necessary to ask: what dividend? It has already been spent. The Associate Minister of National Defence has stated it well in saying "There is no peace dividend to call, and in fact much of the stock that helps produce the dividend has been sold. Our concern will...be to make sure we don't sell it all off".[5]

A military force cannot remain effective without proper funding. Equipment wears out or becomes unsafe, personnel need to be paid, fed and housed. And without maintaining the personnel and the equipment the force can lose its ability to perform its job. As a former Deputy Minister of Defence testified,

> *"Deterrence is achieved by the evident ability to field and sustain effective defence. Without the evident sustainability there is no real deterrence. You are deluding yourself more than you are deluding the other guy. What is required is an ability to sustain the force in being until the nation is on a prepared wartime footing. In my judgment, 180 days would be the absolute minimum that you would need for not only Canadian industry but American industry to really move to anything like a wartime footing. You must be able to survive for 180 days with the forces and the equipment that you have. This means that the 30 days of ammunition and consumables we talk about just last for 30 days. The other 150 days are not covered."* [6]

If a country's armed forces are to do their job effectively they must have adequate funding, predictable funding, long-term funding commitments. The tap cannot be turned on and off at short intervals and the country still expect to have an effective military. It takes 10 years or more to design and build a ship, and longer to develop a ship-building industry that can respond to demand when it is needed. It takes time to train experienced personnel to fly the high-technology aircraft of today. It takes time to develop experienced commanders.

In recent years successive Canadian governments have promised to budget in a way that will permit the military to plan ahead and develop the personnel, equipment and systems that any such organization needs. In the early 1980s we responded to the NATO request that all countries commit to budgets with real growth (after inflation) of 3% per annum. Some years we met it, depending on how creatively the arithmetic was applied, such as whether we counted our contributions to the NATO infrastructure and to the AWACS program (Airborne Early Warning and Control

---

[5] Hon. Mary Collins, delivering the Philip E. Uren Memorial Lecture at Carleton University, April 3, 1990.

[6] C.R. Nixon, addressing the Senate Special Committee on defence. May 31, 1988. p. 10 of Proceedings.

System). Later, the 1987 White Paper adjusted the commitment to "two per cent per year after inflation" for a fifteen year planning period, with "increased resources... necessary in some years as major projects forecast in this White Paper are introduced".[7]

But then, in the April budget of 1989, those commitments went out the window. Major projects, notably the nuclear-powered submarine program, were scrapped; and the budget was reduced by $2.74 billion, with other cuts to follow, making a reduction of $3.4 billion over a five-year period.

When the crisis in the Gulf erupted in August of 1990 the government found funds to permit the emergency upgrading of equipment, especially in the ships and the shipborne aircraft. But this is not forever. The gratitude of a government in peacetime has a lifespan approaching that of a butterfly; and the navy, the army and the air force could soon again find themselves between the rocks of budgeting and the hard places of indifference.

The decision to phase out the Trackers, the short-range surveillance and patrol aircraft, is a good example of where we may have been foolishly short-range in our planning. Although these aircraft are old and were due for re-conditioning they were considered by many to be excellent for the purpose, useful not only for military patrols but also for fishery surveillance and for use in detecting illegal drug trafficking or illegal immigration movements. Present plans are to replace the service in part with privately contracted civil aircraft and in part by diverting the already-overworked long-range patrol Auroras to fill the breach. For the sake of a short-term appearance of making a saving, this plan may end up costing more in the long run. The significance of the Tracker example is that, when squeezed so hard to make highly visible and politically attractive savings, DND had few options left. There has to be a better way.

Of course, Canada has the option of ignoring our equipment manufacturing capability. We could opt for buying everything offshore - the ships, the planes, the tanks and armoured cars, the rifles, the clothing, the trucks. But that would make our financial situation worse than ever, and we would lose another piece of our independence. Also, as described earlier in this volume, we could abandon more of our ability to defend ourselves, and in the process lose even more of our sovereign independence.

The better solution by far is to plan ahead courageously and

---

[7]White Paper p. 67.

intelligently, and then to stick by that plan even when the fickle winds of opinion may temporarily be unsympathetic or unaware of important long-range needs.

> Defence budgets ought to be recognized as having long-term requirements, and should be protected from temporary cuts that would damage long-term plans; funding should be set at attainable levels that will permit the Defence department to develop its capital programs in a manner that is consistent with military considerations, and good budgetary practice, and that may also assist Canadian industry

D. Public Support

No public organization can long survive and remain viable without public support, and this is very true for the armed forces of Canada. By its very success, working within NATO and NORAD for many years, the forces of Canada have brought the country the longest period of peace we have ever known. The need for defence has taken a back seat in the minds of Canadians. Even the day-to-day work of saving lives through Search and Rescue operations - which number some 9,000 every year, of training thousands of Canadians each year in the biggest education complex in the country; even the quiet, professional attention the members of the armed forces devote to developing and keeping their skills - all these go largely unnoticed by Canadians.

Recruiting is down at the military colleges. There are still plenty of recruits, but the applications do not arrive in overwhelming numbers as they once did. Enquiries at the feeder schools, the secondary schools across Canada, reveal that the problem is not one of a bad image of the Canadian Forces: it is a problem of no image at all! The Gulf war created a tempoary increase in interest; but unless our government shows its respect for the long term needs and professionalism of the Canadian Forces the young men and women with the greatest leadership potential will continue to seek their careers elsewhere than in the service of their country.

The closing of Reserve armouries across Canada had the result of removing military uniforms from the public eye. Without a young Canadian on every street in every town Canadian Forces are no longer able to provide that link with the community they once had. No bands to play appealing and patriotic march music, no mess to provide social contact with leading figures of the community and to serve as a repository of history and tradition, no fitness programs, driver training and trades training programs, given in and near the community. The disappearance of these testimonies to Canada's Forces has resulted in ignorance and indifference to defence problems.

In 1989, when the defence budget was so badly slashed, there was no widespread public outcry. Only the defence associations and a few other knowledgable persons protested. The public had become unaware of defence problems; and unaware means unsympathetic. It is ironic that, outside the defence community the people who pay the most attention to defence are the members of the peace and disarmament groups; and their views on the subject tend to be rather one-dimensional. There will never be a constituency for defence unless all manner of Canadians know about defence. And they cannot know about defence unless they have some exposure to defence representatives and can feel some identity with the challenge and the importance of defence to Canada.

National Defence is not without good mechanisms for disseminating information to the public, and the office of the Director General of Public Affairs (formerly DG Info) has frequently won the respect of news media and the public for its quiet efficiency in meeting the everyday demands for sound, factual information. But since the spring of 1989 there seems to have been a lid placed on these mechanisms, a reluctance to allow them to convey the story of defence to Canadians.

In June of 1990 HMCS Provider, a navy supply ship, conducted a dramatic rescue in the South China Sea. A large number of Vietnamese refugees was rescued by the ship and its crew, given medical and other treatment, and delivered safely to the Phillippines. Did the Canadian government tell anyone? The peacekeeping operations that Canadians conduct with other nations every day around the globe were celebrated by the award of the Nobel Peace Prize to the UN in 1988 to recognize those achievements. Does our government care enough to bring this kind of happy news to the attention of its people?

A recent survey performed by National Defence revealed (or confirmed) what some of us have known - that Canadians are woefully ignorant about defence. One of the reasons is the lack of attention paid to the teaching of history in the schools. Another is the lack of links to the community that I have already mentioned.

In 1983 the department recognized this problem and the Minister authorized the creation of an information unit that included a speakers bureau. Interest at the time was running high on subjects such as cruise missile testing, nuclear weapons testing and the roles of defensive alliances. The members of the National Defence Speakers Bureau, with a modest amount of briefing in policy, arms control and related subjects, were able to participate knowledgably in public meetings, to answer questions and, in general, to help ordinary Canadians to understand the issues and to come to their own conclusions. Most of the members of the Bureau were military officers, but a few were civilians, and some were not part of DND at all, but interested in assisting nonetheless. Most of the speaking appearances were to service clubs, schools, associations and community groups meeting in church basements or town halls.

The National Defence Speakers Bureau operated as a low-key response to public requests for information, and in every case the appearance of these knowledgable representatives, with their genuine commitment to peace and freedom, won respect from their audiences. It is unfortunate that, following the 1989 budget disappointment, National Defence allowed the Bureau to fall into disuse. If members of Canada's Armed Forces and their civilian colleagues and supporters do not make contact with the public, there can be little hope for continuing public support.

There is no lack of good messages for defence reps to convey. Canada's policies are good: we stand for peace, freedom, human rights and the dignity of the person; and the reason for the existence of Canada's defence capability is not so much the defence of land as it is of our values and our hard-won rights.

National Defence would be well advised to resume encouraging its representatives to meet Canadians and discuss defence with them.

A new defence policy should be produced immediately, and updated annually.

Government should recognize the distinctive and special needs of Canadian Forces and assume more responsibility for protecting them from politically motivated pressures that could risk impairing the ability to perform their job.

Defence budgets ought to be recognized as having long-term requirements, and should be protected from temporary cuts that could damage long-term plans and cause long-term waste of funds; funding should be set at attainable levels that will permit the Defence department to develop its capital programs in a manner consistent with military considerations and good budgetary practice and that may also assist Canadian industry.

National Defence should resume encouraging its representatives, members of the National Defence Speakers Bureau, to meet Canadians and discuss defence with them.

The government should take opportunities to publicize the activities and the special achievements of the Canadian Forces.

The government should encourage means of improving the quantity of history taught in Canadian schools.

VI INDUSTRIAL SUPPORT

A. The Essential Foundation

When they think about them at all, Canadians often take their military suppliers for granted. We have geared up for earlier wars from a standing start; why should it be any different now? But there is a difference: today's sophisticated weapons systems cannot be produced quickly. Therefore, even in the era of glasnost and perestroika, prudent nations are careful to ensure their industries are able — at a minimum — to meet the peacetime needs of their armed forces. This, in turn, provides the base for rapid expansion should it be demanded in time of war.

The ability of Canadian industry to support the country's military is an essential component of national security. The ability to sustain armed forces during combat is a major factor in deterring an adversary. If Canada has to rely on other countries to supply its defence needs, it becomes vulnerable to a disruption of that supply in wartime. That disruption could be caused either by a foreign government's redirection of defence production to its own wartime needs, or by enemy action.

However, as defence specialist Dr. David Haglund has noted, "if it can be said that dependence in peacetime can become a dangerous vulnerability in wartime, so too must it be noted that fostering inefficient defence industries through protectionism can also take its toll on the national security."[1]

While it would be unrealistic to expect Canadian industry to meet all the military's needs, it is not demanding too much to expect the government to implement a security policy that encourages defence production in certain key areas. But any defence industrial policy must, however, be developed with Dr. Haglund's warning in mind.

B. The Major Industrial Sectors

Canada's armed forces are relatively small. They are also assigned a variety of domestic and international roles. Consequently, the military requires small numbers of a large variety of equipment. Canada's defence industries have been unable

[1] David G. Haglund, "Introduction", in David G. Haglund, ed., *Canada's Defence Industrial Base*, (Kingston: Ronald P. Frye & Company, 1988), p. 3.

to survive on such a limited home market and have thus turned their attention to exports. This has resulted in a defence industry which is more attuned to the military needs of foreign governments and which has become more and more specialized in producing sub-systems and components for offshore prime contractors, rather than complete weapons systems.

1. Aerospace Industry [2]

The Canadian aerospace industry employed approximately 60,000 people in 1989 and today has annual sales of between $8 and 9 billion, of which just over 30 per cent are defence related. Exports account for about 70 per cent of total sales. Ninety per cent of the industry is located in Ontario and Quebec, 4 per cent in the maritimes, and 6 per cent in western Canada. The industry's major customers are U.S. prime contractors.[3]

The four largest aerospace firms — Pratt & Whitney Canada Inc., Canadair Inc., The de Havilland Aircraft Company of Canada Limited and Bell Helicopter — account for 45 per cent of the industry's sales. These companies are capable of designing, developing, manufacturing, marketing, and repairing complete aircraft, engines and systems.[4]

Another 40 medium-sized companies — such as McDonnell Douglas Canada Ltd., Fleet Aerospace, Spar Aerospace Limited — account for another 45 per cent of the industry's sales. The companies generally supply other prime aerospace manufacturers with proprietary products or build-to-print manufactured components. They are also capable of repairing and overhauling aircraft, engines, and components. These companies have annual sales ranging between $20 million and $400 million.

The remaining 10 per cent of the industry's sales is contributed by the approximately 140 small special process and precision machining businesses which survive mainly on short-term orders from large companies, aerospace parts distributors and foundries. These small companies generally have annual sales of less than $20 million.

---

[2]The division of defence industries into aerospace, defence electronics, shipbuilding and ship repair, automotives and munitions, is arbitrary. Some companies are involved in more than one sector and thus their production figures may be double-counted. Despite this caveat, the information provided is enough to give the general reader an overview of Canada's major defence sectors.

[3]Aerospace Industries Association of Canada figures.

[4]Industry, Science and Technology Canada, Industry Profile: Aerospace, (Ottawa: 1988). Most of the information in this section came from this report.

The larger companies are mostly foreign-owned, while the smaller companies are generally Canadian-owned. The industry is highly specialized, and there is only limited direct competition among the major companies.

Within the aerospace industry there are approximately 50 companies involved in space-related work. In 1987, these companies generated $400 million in sales (70 per cent of which was exported) and employed 3700 people. Spar Aerospace Limited is the largest firm and accounts for about one-half of total sales and employment. It is the only company capable of assuming the role of a prime contractor and of manufacturing complete systems. Other firms are relatively small, with annual sales between $20 million and $30 million. Most of the companies are Canadian-owned.[5]

Aerospace sales have continued to climb over the past decade because of airline deregulation and the growth in the small turboprop and turbofan aircraft commuter and corporate market. The industry has survived the ups and downs of defence procurement because it has proved flexible by diversifying into civilian products and export markets.

The industry pumps about 10 per cent of its sales back into research and development — a smaller percentage than the United States, which devotes 17.5 per cent to R&D.

Current products of the industry include: reconnaissance drones, turboprop/turboshaft engines, flight simulators, undercarriages, transport aircraft and light helicopters.

2. Defence Electronics

The Canadian defence electronics industry, centered mainly in Ontario and Quebec, employs approximately 26,000 people and has annual sales estimated at $2.4 billion. Exports, which are mainly to the United States, account for 80 per cent of the industry's sales.

The industry is dominated by 10 major companies: Litton Systems Canada Limited, CAE Electronics Ltd., Canadian Marconi Company, Unisys Canada Inc., Raytheon Canada Limited, Computing Devices Company, ITT Canon Canada, Allied-Signal Aerospace Canada Inc., Rockwell International of Canada Ltd., and MacDonald Dettwiler and Associates Ltd.[6]

---

[5]Industry, Science and Technology, *Industry Profile: Space*, (Ottawa: 1988).

[6]Industry, Science and Technology Canada, *Industry Profile: Defence Electronics*, (Ottawa: 1988), p. 1 (with amendments by R. Hicks).

As in the case of the aerospace industry, most of these major companies are foreign owned. Smaller defence electronics companies tend to be Canadian-owned. The companies, on average, devote 13 per cent of their sales revenue to research and development.[7]

All the companies generally produce sub-systems for prime contractors or the government. Most Canadian companies are too small to support the investment required to be self-reliant in key components — such as Very High-Speed Integrated Circuits (VHSIC). They have to be imported. Because the next generation of defence electronics products is based on these components, Canadian industry will find itself at a disadvantage in the future development of new technologies.[8]

One evolving area in which Canadian companies are fighting to become proficient, is systems integration. To date, Canadian industry has limited capabilities but with technology transfer from the United States, this may become a profitable area.[9]

The industry's products include: ballistic computers for main battle tank fire control systems, land vehicle navigation systems, communications equipment, night observation devices, gun alignment and control systems, and portable artillery computers.

3. Shipbuilding and Ship Repair

The shipbuilding and ship repair industry in Canada has been in decline for a number of years. Employment has fallen from 16,000 in 1975 to 8,600 in 1989. The value of new vessel construction and repair work in 1989 was approximately $605 million, of which only 3 per cent was for export — and that 3% was all in the repair area, not new construction. Government orders currently account for over 90 per cent of new construction.[10]

The industry, which includes about 53 yards, is largely Canadian-owned. MIL Group in Quebec and Saint John Shipbuilding Limited in New Brunswick account for nearly 50 per cent of the total shipbuilding capacity. Virtually all of the major

---

[7]Ibid., pp. 1-2.

[8]Ibid., pp. 3-4.

[9]Ibid., p.4. With the New Shipborne Aircraft program, Westland and Agusta (owners of E.H.I. Canada Inc. the prime contractor) found they were unable to integrate the various missions systems with the aircraft. They brought in U.S. giant Unisys Corp., parent company of Paramax Electronics Inc. in Montreal, as an equal partner in EHIC in order to share the financial risk and help shoulder the technological burden.

[10]Figures from the Canadian Maritime Industries Association.

yards construct a broad range of commercial vessels, but MIL, Saint John, and Versatile Pacific Shipyards Inc. on the west coast have the capability to construct naval vessels.[11]

The Canadian industry is characterized by a significant amount of unused production capacity, a small domestic market, high

HMCS Halifax under construction at an East coast shipyard.

[11]Industry, Science and Technology, *Industry Profile: Shipbuilding and Ship Repair*, (Ottawa: 1988). Most of the information in this section comes from this report.

wage rates, occupational trade demarcation practices that hinder productivity, and outdated equipment which cannot accommodate new types of assembly-line production.

Canada has been handicapped in both the export and the domestic markets because of international overcapacity and the entry into the marketplace of low-cost competitors such as Korea, China and Taiwan. The Canadian industry has thus been virtually shut out of international markets, except for some small amounts of minor structural work and ship repair from the United States and emergency repairs.

To become more competitive, the industry has begun the process of rationalization and modernization. But its prospects internationally remain limited. One major problem is that other countries aid their industry by providing subsidies and favourable financing terms; Canada does not.

The Canadian industry has been hit hard domestically by the cancellation of the nuclear-powered submarine and the Polar 8 icebreaker programs. Current naval programs include the 12 anti-submarine warfare frigates, the modernization and reconfiguration of four Tribal class destroyers, and the beginning of a program to design and construct 12 Maritime Coastal Defence Vessels (mine countermeasures vessels). Beyond these programs, which will be completed in the late 1990s, there are no new naval projects.

4. Automotives

The automotive industry employs approximately 154,000. In 1988, the value of its shipments was $42 billion.[12] Defence economist John Treddenick found that for 1984-85 only 2 per cent of the industry's employment and shipments were defence-related.[13]

Few of the high-tech components for military trucks are produced in Canada. Instead the automotive industry relies on the import of parts which can be assembled in Canada. The extensive foreign ownership and low volume of production have inhibited both research and development and the adoption of more efficient and innovative production technologies.[14]

---

[12] Industry, Science and Technology, *Statistical Review of the Canadian Automotive Industry: 1988*, (Ottawa: 1989) pp. 18 and 38.

[13] John Treddenick, "The Economic Significance of the Canadian Defence Industrial Base", in David G. Haglund, ed., *Canada's Defence Industrial Base*, (Kingston: Ronald P. Frye & Company, 1988).

[14] Industry, Science and Technology Canada, *Industry Profile: On-and Off-highway Medium/Heavy-duty Trucks*, (Ottawa: 1988).

Most of Canadian production of automotives for the Canadian Forces is done under licensing agreements with foreign countries.

The Diesel Division of General Motors of Canada in London, Ontario, is a major producer of wheeled Light Armoured Vehicles. The company began producing the military vehicles, under licence from Motorwagen Fabrik AG of Switzerland, after receiving a contract in 1977 for 491 general-purpose armoured vehicles for the Canadian Forces. It has gone on to develop more than nine variants and to secure substantial orders from the U.S. Marine Corps and U.S. Air Force.

Bombardier Inc. produced both the German Iltis light utility vehicle and the American M35 medium logistic vehicle under licence for the Canadian Forces. Invar Manufacturing has the contract to manufacture TOW turrets for the Canadian army under licence from Thune-Eureka of Norway. And in 1988 Urban Transportation Development Corp. of Kingston won the $250 million contract to produce, under licence from Steyr-Daimler-Puch of Austria, 1,122 heavy military trucks for the Canadian army.

5. Munitions

The munitions industry consists of three prime contractors, their sub-contractors, and small specialty manufacturers. Its annual sales are approximately $1 billion of which about $600 million worth are defence-related. About 50 per cent of the defence production is exported.[15]

This is the one sector in which the government has intervened to ensure Canada has a production capability that meets the military's needs.

In 1978 the government approved a Munitions Supply Program to establish and maintain a Canadian industrial capability for the production of high volume usage ammunition and associated stores.

Although the government provided limited funding to modernize plants, many of the facilities are still of World War II vintage. Also, the companies still rely on foreign suppliers for some key components so that the major Canadian producer of ammunition is primarily a "load, assemble and pack" facility, according to Canada's Auditor General.[16]

Approximately 75 per cent of DND's annual procurement con-

[15] Supply and Services Canada, *The Defence Industrial Base Review 1987*, (Ottawa 1987) p. 5.
[16] *Report of the Auditor General of Canada to the House of Commons*, Fiscal Year Ended 31 March 1988, sections 15.17-40.

tracts for ammunition are let to Canadian companies, but that ammunition costs about 30 per cent more than the lowest prices available from other NATO suppliers. Also, in an effort to sustain the industry, the Auditor General found DND has been purchasing more ammunition than necessary and this resulted in increased storage costs for the Department.[17]

SNC Group Inc. of Montreal is the largest producer of ammunition in Canada. It owns both IVI Inc. and Canadian Arsenals Ltd., a former Crown corporation. Recent cutbacks in DND's ammunition procurement have caused some problems for the Montreal-based company.

In 1984 DND supported the establishment of a small arms industry by buying the licence and plant equipment for the Colt M-16 rifle. It then supplied Diemaco of Kitchener, Ontario with the licence and the machinery to produce 81,500 of the weapons for the Canadian Forces.[18]

C. The Market and Government support

1. The Market

In the late 1950s the government realized that Canadian industry could no longer competitively develop and produce major weapons systems. So the government entered into a series of arrangements — collectively known as the Defence Production Sharing Arrangements — that would increase the flow of defence supplies between Canada and the United States, while at the same time giving Canadian companies access to the large American market.

That access, combined with the shifting fortunes of Canadian defence procurement plans, has led Canadian defence companies to rely on exports as the mainstay of their survival.

But this can be carried too far. A 1988 report by the Department of Supply and Services revealed, "A serious concern [in the munitions industry] is the extent to which normal peacetime production is exported to our allies, particularly the United States. In any emergency it may not be possible to divert the normally exported production to meet Canadian needs, due to planned agreements with, in this case, the United States Department of Defense."[19]

---

[17]Ibid.
[18]"New Approach to Canadian Defence Procurement", in *Jane's Defence Weekly*, November 21, 1987, pp. 1200-1201.
[19]Supply and Services Canada, *The Defence Industrial Base Review 1987*, (Ottawa, 1987) p.5

The U.S.-Canada defence trade is not without its other problems as well. Increased lobbying from pressure groups in the United States has resulted in a number of non-tariff barriers in the Defence Production Sharing Arrangements. For example, there are laws preventing Canadian suppliers from bidding on American contracts for naval vessels or for certain articles containing a number of specialty metals. Also, there are set-asides for small businesses, labour-surplus areas, minority groups, and depressed industries. Canadian companies are also handicapped by national security considerations which limit access to sensitive information.[20]

Canada has signed development and production sharing agreements with other NATO countries, but the two-way trade is much smaller. Between 1959 and 1989, Canada exported $15.4 billion worth of defence goods to the United States, compared to $5.4 billion worth to all other countries. In one year, 1989, the United States received over 80 percent of Canada's defence exports.[21]

Wherever Canada's defence industry looks for export markets, non-tariff barriers for high-technology products remain a problem. Consequently, the government should play a lead role in helping improve market access by governmental links, or by paving the way for joint ventures.

2. Government Support For Defence Industry

The Canadian defence industrial base accounts for less than 1 per cent of the Gross Domestic Product and less than 1 per cent of Canada's total merchandise exports.[22] Moreover, Canadian defence industries employ only 85,000 people directly and 65,000 indirectly — less than 1 per cent of total employment.[23] Obviously, the economic importance of a defence industry is not great when measured in terms of dollars and employment. But are there other reasons for creating and maintaining a defence industrial base?

Although the industries' impact on the economy as a whole may be considered insignificant, certain industries such as shipbuilding and to a lesser extent, aerospace, depend on defence production

---

[20] Robert Van Steenburg, "An Analysis of Canadian-American Defence Economic Cooperation: The History and Current Issues", in David G. Haglund, ed., *Canada's Defence Industrial Base*, (Kingston: Ronald P. Frye & Company, 1988), pp. 204-207.

[21] Department of External Affairs figures.

[22] Treddeneck, pp. 42-43.

[23] L. John Leggatt, "Technological Innovation in Canadian Defence Industry", in David G. Haglund, ed., *Canada's Defence Industrial Base*, (Kingston: Ronald P. Frye & Company, 1988), p. 57

for their existence. As well, some defence companies are very important to the local economy in which they operate. Also, because defence industries tend to be high-technology industries they can act as a stimulant to the economy as a whole.

But the most important reason is the one stated at the outset: the maintenance of an industry capable of supporting a country's armed forces, not only in peacetime, but also during the surge production rates of wartime, acts as a deterrent to a potential adversary.

Despite the vital part a defence industrial base plays in the deterrence of war, government has not produced an overall national security policy which includes a defence industrial strategy. It is time to do so.

European governments take an active role in shaping their defence industrial base. For example, West Germany has designated two companies as military computer suppliers. Any research and development or production to be done will be done by one of those companies. Because they are still in competition they have an incentive to pursue new technologies and keep costs low.[24]

The Swedish industry's success is due to its stability. That stability stems from a five-year defence budget plan which allows industry to make long-term equipment and manpower commitments, and from a planned phasing in and out of major weapons systems, keeping production continuous and smoothing the transition from one generation of equipment to another. The Swedish defence industry is also well-integrated with the civilian industry because of a law that forbids any company to have more than 25 per cent of its business in defence. This allows the companies to survive during periods of decreased defence spending and makes them more cost-sensitive.[25]

The United States does not have a defence industrial strategy, but defence industry expert Jacques Gansler has recommended that Washington implement one based on five considerations:

1. A research and development investment strategy. He says the Defense Department "must identify those specific long-term technologies that should be pursued and, with a conscious effort, fund multiple firms to pursue these areas."

---

[24]Jacques S. Gansler, *The Defense Industry*, (Cambridge, Mass.: The MIT Press, 1980), p. 245.
[25]Ibid., p. 246.

2. Creation of incentives for productivity gains. Gansler suggest the Defense Department should make greater use of the "design-to-cost" concept — the specifications for a piece of equipment should be the best system for a specific price. "Using design-to-cost, engineers are forced to address production considerations, as well as the cost of the new materials and technologies that they apply, right up front in their very-preliminary designs."

3. Far greater integration of civil and military production. Military production costs are driven higher because of demands that industry follow military specifications and its specialized "way of doing business". Gansler says, "Even for those cases where the DoD could not use commercial equipment, they could still realize many of the same benefits if they encouraged the integration of the production facilities — so that, for example, low-volume military electronics would be co-produced with high-volume commercial electronics. This could have the added advantage of providing production surge capability in times of military crisis."

4. Implementation of a defence industrial strategy through major weapon system and budget decisions. By giving consideration to which, when, and where, weapons systems are to be built, which facilities to invest in, and by planning production so that it is sequential and avoids bottlenecks, the Defense Department would actually be implementing an industrial strategy rather than making a series of decisions on a program-to-program basis.

5. Making defence industrial strategy part of U.S. national security strategy. "Essentially, this means viewing industrial efficiency and crisis responsiveness as being in the same class as strategic and tactical forces — as a means of deterring potential adversaries".[26]

Dr. Gansler's policy plan for the much larger United States defence industry is not totally applicable to the Canadian scene. But his suggestions do point out a sensible policy route for Canada. Because of DND's equipment requirements — both ini-

---

[26] Jacques S. Gansler, "Needed: A U.S. Defense Industrial Strategy", in *International Security*: Fall 1987, pp. 45-62.

tial procurement and subsequent life-cycle support — the first priority of a defence industrial base policy should be to meet that department's needs. Therefore, DND should be designated as the lead department for producing and implementing such a policy.

At the present time, the responsibility for nurturing a defence industrial base is spread among several government departments. The Aerospace Industries Association of Canada told a Senate Committee in March 1984:

> *"The linkage between defence policy and the defence base has almost been lost. DND has a need for industrial support. However, it has no mandate. The mandate, the Defence Production Act, is in Supply and Services. They have no money. Moreover, it is inconsistent with their service agency role. The funds are with Regional Industrial Expansion [now the Department of Industry, Science and Technology] under the mantle of industrial support programs. These programs, however, are influenced more by regional and export factors than by defence considerations as they very well should be. In addition, the policy framework within which defence-oriented support programs operate, resides in External Affairs ...."* [27]

The next step must be to establish a process which allows DND and industry to consult on the military's long-term requirements. These deliberations should include: what needs can be met by Canadian industry; what requirements would require a special R&D effort to be met by Canadian industry; and what items would have to be purchased offshore.

Access to long-term military planning is essential if industry is to have the time to acquire the necessary machinery and tooling for long lead time components and systems.

To its credit, the government has established a Defence Industrial Preparedness Advisory Committee (DIPAC) made up of academics and industry representatives to advise the Minister of National Defence on national defence industry issues. Its job is to identify industrial deficiencies that need to by addressed by DND, and to suggest areas where DND may want to concentrate its efforts.[28] After a rather slow start, the Committee now appears ready to fulfill its mandate.

[27]Senate Special Committee on National Defence, *Minutes*, March 14, 1984, Issue No. 4, p. 4A: 13.
[28]*The Wednesday Report*, August 10, 1988, p. 3.

At an international (or rather, bilateral) level, there is the North American Defence Industrial Base Organization (NADIBO). Established in 1987, NADIBO has been actively working on production base assessments based on Canadian and American military requirements and industrial capabilities.

Both DIPAC and NADIBO have the potential to assist in the establishment of an effective and efficient defence industrial base. But that potential will only be fulfilled if the government listens to their recommendations and acts accordingly. Procurement decisions should be used to shape the defence industrial base.

The government should gear its procurement decisions to take advantage of industry's strengths and rectify its weaknesses based on DIPAC's and NADIBO's assessments, rather than on any short-term political advantage.

There are four methods of military procurement, as listed by the late Dr. Rod Byers, in his report for the MacDonald Royal Commission: national procurement; licenced production; offshore purchases; and joint ventures.[29] All methods have their advantages and disadvantages, so the government should use a combination to aid the development of an adequate defence industrial base.

National procurement requires that industry have research, development, testing and production capabilities. Any weapons system procured domestically should preferably have a long production run and export opportunities to offset the high start-up costs.[30]

Identifying export markets is central to expanding Canada's defence industrial base. If a company is able to find offshore customers for its product, then it will survive in peacetime, and have the needed excess capability for wartime. DND can help in this process by keeping to a minimum any characteristics that would make a weapons system unsaleable in the global marketplace. A good example of a company being stymied in its search for export markets is Litton Systems Canada Limited. Litton is the prime contractor and overall integrator for the Tribal class destroyer update and modernization (TRUMP) program. Normally a company undertaking such a project would have hopes of selling its new products abroad but in Litton's case, its export prospects are severely limited. The software it had to develop is so TRUMP-specific, it is unlikely that anyone else could use it.

Licensed production can benefit industry because it usually involves a technology transfer from the prime contractor. This in

---

[29] R.B. Byers, "Canadian Defence and Defence Procurement: Implications for Economic Policy", in Denis Stairs and Gilbert R. Winham, Research Coordinators, *Selected Problems in Formulating Foreign Economic Policy*, (Toronto: University of Toronto Press, 1985), p. 153.
[30] Ibid

turn can lead to export sales or a life-cycle support capability. Licensed production, however, does not involve research and development, and quite it often leaves Canadian industry just putting together imported components.

Procurement from offshore sources can only benefit Canadian industry if the government negotiates a package of industrial benefits which include defence business and/or technology transfers. After a few initial defence deals showed flaws in the benefits approach — a $2.9 billion offset package obtained from McDonnell Douglas when the government bought the CF-18 fighter included such things as the enhancement of tourism and export marketing assistance; only 10 per cent of the offsets were tied to technology transfer — the government placed more emphasis on technology transfer and co-operative development and production.

The obvious benefit of offshore procurement is that it is usually cheaper than using a Canadian source (especially if the contract was to involve establishing a Canadian source). This frees up defence money for other capital programs which can be sourced domestically.

Because offshore procurement makes DND dependent on offshore suppliers, such purchases should be accompanied by an effort to ensure that life-cycle support is established with Canadian industry. DND did this with the CF-18. Canadair was awarded the support contract for the aircraft and received the necessary technology from McDonnell Douglas.

Joint ventures benefit Canadian industry by spreading the research and development costs among several participants. This allows all those involved to be in at the critical front-end of the design process, while substantially reducing the financial risk. Joint ventures also benefit the participants by sharing the expertise. Joint ventures, however, also require a high degree of international co-operation, from design considerations to financing strategy. And as the demise of the 8-nation NATO Frigate Replacement program demonstrated, international projects should be limited to no more than 3 or 4 participants.[31]

Government must devote more money and effort to research and development. Defence industries are high-technology industries and exist in a highly competitive marketplace. Research and development is their lifeblood.

---

[31]For a good analysis of what went wrong on the NFR program, see Daniel Hayward, "The NATO Frigate Requirement for the 1990s: A Case Study in Multinational Cooperation in Defence Production", Paper Presented at the U.S. Naval Academy Foreign Affairs Conference, April 9-12, 1990.

DND carries out research and development at its six research establishments and by contracting out to industry, universities, and other government departments. The department targets 5 per cent of its capital budget for R&D. In 1989-90, the total R&D budget was $254 million, of which $91 million went to industry. In 1990-91, $269 million was spent on the R&D program, $93 million went to industry.

Those figures include the funds for the new Defence Industrial Research Program (DIR) announced by former Defence Minister Perrin Beatty in the spring of 1988. The DIR program was initially funded at $5 million, and the intent is to raise that to $15 million annually over time. The program, which is a 50:50 cost-sharing one between DND and industry, was established to encourage technology transfer from departmental laboratories to industry and to improve the position of Canadian companies in supplying high-technology equipment to the Canadian Forces and its allies.[32]

Based on a random survey of defence companies, technology expert John Leggatt has identified two important impediments to increased technological innovation in the defence industrial base: insufficient funds for innovation, and government procurement policies. He suggests that "a closer alignment is needed among the long-term needs of DND, corporate capabilities, priorities for contracted research, DND laboratory applied research, and DND development programs."[33]

He does not believe that the way to increase Canada's defence industrial base is through subsidies, but rather through help in research and development. He says, "Subsidization through preferred procurement practice reduces the effectiveness of the defence dollar in providing for needed defence equipment, and ultimately could lead to a deterioration of defence industry competitiveness. Rather than paying extra for Canadian-produced goods, DND should provide for increased funding and co-ordination at the beginning of the innovative process."[34]

The government's Defence Industries Productivity Program (DIPP) does provide some money for research and development. The program had its origins in 1959 when the DPSA was concluded. DIPP funds are used to sustain industry's technological capabilities for the purpose of generating exports. In 1990-91, the government allocated $236 million to this program.

[32]*National Defence, 1990-91 Estimates: Part III - Expenditure Plan*, p. 141.
[33]Legatt, pp. 67-68.
[34]Ibid., p. 68.

But because DIPP funds emphasize products for export markets, the money is being used for products which may not coincide with DND's needs. Industry has also complained that since DIPP is administered by the Department of Industry, Science and Technology (previously the Department of Regional Industrial Expansion), it has become an instrument for regional enhancement rather than defence industrial enhancement.

Finally, no defence industrial base policy would be complete without provision for the training of new workers. The government must take the initiative in establishing apprentice-training programs to support the defence industry.

John Leggatt's survey of defence companies revealed that one of the main obstacles to innovation was the limited availability of talent. "The industries reporting this either use very specialized technology, and so must train engineers and scientists in-house, or have little human resources available to devote to developing ideas for new products."[35]

More emphasis on the sciences in our educational system and government financial aid for apprenticeship programs would go a long way to helping solve the problem.

---

[35]Legatt, p. 60.

SUMMARY OF RECOMMENDATIONS

So, what lessons are Canadians to take from the various debates about defence that have been taking place lately? One place in which they might look is into the context of force structures. In an exchange of letters in April of 1991 General A.J.G.D de Chastelain, Chief of the Defence Staff, spelled out his view of the need to keep a balanced force of navy, army and air forces, including an expeditionary capability with which to meet overseas emergencies. By contrast, the retiring Vice Chief, Vice-Admiral C.M. Thomas, maintained that budgetary limitations make that plan unobtainable, and instead advocated concentration on the equipment needs of the navy and the air force, if necessary at the expense of the expeditionary capability. Both of these very experienced leaders agree that we have too many military bases for the budgets of the 1990s, and these bases consume funds that might be better used for paying the personnel and for updating the equipment on which they must rely in order to do their jobs safely.

And what lessons have we learned from the recent war in the Persian Gulf? Perhaps the first and most obvious lesson is that Kuwait was insufficiently defended; it lacked deterrent capability and so was invaded by the aggressor. But there is more: I would like once again to quote the words of C.R.(Buzz) Nixon, who had this to say in a recent issue of National Network News - a publication of the Defence Associations National Network:

> " *The Gulf war has once again shown that people who value freedom, democracy, the rule of law, and respect for the institutions and processes of free societies that flow from these concepts must not forget that they have not been obtained and are not retained at zero cost. As long as tyrants like Saddam Hussein exist, maintaining these pillars of our democratic societies will require unrelenting vigilance and, in the extreme, a willingness to use the force of arms.*

> *The war also demonstrated that the world has become a "global village". Apparently local or regional actions of individual countries can and do have world wide consequences. Stability and order in this "village" require strengthening of the capabilities of the United Nations to act as the internationally accepted institution responsible to establish the norms of world order, and to have the capability to maintain that order.*
>
> *For the United Nations to have the enforcement capability to establish and maintain world order means there is going to have to be some international arrangement by which the UN will be able to assemble at short notice, in accordance with previously established procedures, the military forces required to give effect to UN resolutions. Leading democratic nations, which includes Canada, should be the leaders both in developing the procedures for the establishment of such forces, and for structuring their national defence activities so as to be able to contribute meaningful and effective military capability when called to do so.*
>
> *...If Canadians believe that a stable world order is to our own and to mankind's benefit, that the UN should increasingly play a larger role in world affairs, that "forces in being" are essential for the UN to be effective, then Canada has no option but to ensure that its military capability is commensurate with Canada's place in world affairs."*

Lastly, it is worth remembering the massive displays of public support for the Canadian Forces personnel who left Halifax in August of 1990 to enforce the UN resolutions in the Gulf. Commodore Ken Summers, who led the Task Force to the Persian Gulf, has described the hard-working professionalism of the crews, both military and civilian, who worked against time to re-equip the three ships in August, the sad but proud farewells from Halifax, the very high standards of military and personal conduct while the Canadians were in the Gulf area, the gifts received from family and friends - and even from perfect strangers, and then the tumultuous welcomes for the safe return home. Whatever their feelings about the need for military action, it was plain that Canadians supported their sailors, soldiers and fliers.

It is up to Canada to ensure that there will always be those sailors, soldiers and fliers, suitably equipped and suitably trained, whenever they should be needed again.

BROAD GOVERNMENT RESPONSIBILITIES

Canada must continue to be a proud member of NATO. The expression that membership may take, especially in its military dimension, is another matter and one that offers opportunity for constructive debate.

Serious thought should be given to creating either a pair of broad, non-partisan Councils to analyse and digest advice from many quarters, and to advise Cabinet as required – one for the review of foreign and security policy and one for defence policy – or to a single Security Council with separate sub-committees for foreign and defence policy. Members should represent a variety of related experience and a mix of the political and non-political.

The Canadian Security Policy and the Canadian Defence Policy in Canada's interests should be produced immediately, and update annually.

Government should recognize the distinctive and special needs of Canadian Forces and assume more responsibility for protecting them from politically motivated pressures that could risk impairing their ability to perform their job.

Defence budgets ought to be recognized as having long-term requirements, and should be protected from temporary cuts that could damage long-term plans and cause long-term waste of funds; funding should be set at attainable levels that will permit the Defence department to develop its capital programs in a manner consistent with military considerations and good budgetary practice that may also assist Canadian industry.

Improve the efficiency and cost-effectiveness of Canadian air surveillance services by clarifying both the separate and shared responsibilities of the Ministry of Transport and National Defence, by amending the Aeronautical Act of Canada. Canada should contribute joint surveillance information to NORAD.

Extend the ground-based surveillance of Canada's Northern and Arctic regions to cover all the territory claimed by Canada.

Prepare a National Air Surveillance and Control Plan for Canada, with the mandate that it include all Canadian air space.

National Defence should resume encouraging its representatives, members of the National Defence Speakers Bureau, to meet Canadians and discuss defence with them.

The government should take opportunities to publicize the activities and the special achievements of the Canadian Forces.

The government should encourage means of improving the quantity and quality of history taught in Canadian schools.

MARITIME FORCES

The government should not discount the possibility of acquiring a nuclear-powered submarine fleet at some future time.

The two batches of 12 frigates that are now on stream should be completed and put into service as soon as possible.

The program for acquisition of Maritime Coastal Defence Vessels should proceed as rapidly as possible.

The program for replacement of the Sea King shipborne helicopters with the new EH 101 should proceed as rapidly as possible.

Replace the Tracker patrol aircraft, for inshore surveillance, for fisheries patrol, drug interdiction and detection of illegal immigration with aircraft suitably fitted for the the role.

Since the number of fighting ships we are keeping at present is approximately 20, and since the life expectancy of such a ship is 20 years, a way should be found to award contracts for the construction of one such ship each year, thus keeping the art alive in the shipyards, as well as providing a predictable and assured supply of replacement vessels for the users. Canada should explore the possibility of marketing frigates to allied countries.

Participation in STANAVFORLANT is beneficial for Canada as well as for NATO and should continue.

Recognition of the fact that Canada has no maritime capability for establishing surveillance and control of our sea and undersea territories, i.e. for exerting Canadian sovereignty in those areas we claim as Canadians. Maritime aircraft are no substitute for undersea surveillance in ice-filled waters.

The time has come for Canada to consider how best to reorder the powers and resources for surveillance and control of our maritime sovereignty. More efficient use should be made of the military and civilian resources now available. We could also hope for a system that offers faster response to emergencies at sea.

Canada should consider a better co-ordination and consolidation of its maritime responsibilities, for greater effectiveness and possible savings in managing its ocean resources. Placing all non-naval responsibilities under one Minister would be one worthwhile way of approaching it.

For better coordination of assistance to civil authorities a common operations centre, similar to that of the Search and Rescue operation, should be considered.

A decision should be made immediately to proceed with a program of conventionally-powered submarines to replace the Oberons. The first submarines of a minimum six-boat program should be of the diesel-electric type, but the technology for an air-independent propulsion system should be studied and evaluated. If that technology matures over the next five years the Department of National Defence should be ready to incorporate it into the remaining boats, and as a retrofit plug into the first subs off the production line.

LAND FORCES

Seek permission from NATO to re-define and re-organize our European commitment as follows:

Restore the CAST commitment to reinforce North Norway in time of crisis.

Increase airlift capability, to improve the CAST capability.

Commit 4CMBG to the CAST responsibility, but base it in Canada.

Reduce the permanent strength at CFB Lahr to that of a staging base, with resources to support units that arrive there for training and exercises; or better still, strike a deal with the Germans to trade the Lahr base for a staging base in northern Germany, for the rapid reinforcement of Norway, in exercises and in a time of emergency.

Sell off all unnecessary infrastructure at Lahr.

Make greater use of pre-positioned equipment, including the use of allied equimpment, and ensure that Canadian troops participate, on a fly-in basis where required, in appropriate exercises with NATO allies, either in Norway or on the Central Front.

Buy no more tanks at present.

Continue participation in NATO exercises in Europe.

AIR FORCES

Deploy the CF-18s in the defence of Canada and North America, and for rapid reinforcement of NATO forces in North Norway.

Replace and augment the transport aircraft, both fixed wing and helicopters, to enable an expanded capability to airlift troops for exercises, and to support NATO roles, peacekeeping and other UN and humanitarian activities.

Given the need for a strong air surveillance capability for Canadian coastal and open waters, consideration should be given to doubling the number of Aurora or similar long range patrol aircraft from the present 18, as recommended by the Senate sub-committee on National Defence in 1983.

Reinstate the program to modify all seven Challengers for electronic warfare training.

Acquire no additional CF-18s in the near future.

Continue the process, already begun by Air Command, of rationalizing the fleet by reducing the variety of aircraft.

Continue participation in NATO exercises, including rapid reinforcement exercises, and in NORAD exercises.

The Total Force Concept ought to be continued, while allowing for specialized roles such as Naval Control of Shipping and Mine Countermeasures for naval Reserves and Land force units concentrating on northern training, anti-terrorism and peacekeeping.

DND should ensure that Reserve activities and Cadet activities are publicized suitably.

RESERVES

The White Paper goal of 90,000 Reserves should be restored, beginning with a drive to raise the militia to 30,000 from its present 20,105.

Increase the size of the militia and assign it specific tasks, with emphasis on territorial defence and on peacekeeping.

Reserve facilities should be re-opened where necessary, including those new growth parts of Canada that have none, in order to allow a military presence and to establish links with those communities. Facilities need not be lavish.

The University officer training programs (UNTD, COTC, URTF) should be restored and supported.

The program to arrange for Reserve personnel employed in the private sector to be allowed leave for military training purposes should be vigorously pursued.

The process of making the pay and basic pension, disability and death benefits of the Reserves comparable to those of the Regular Force should be quickly concluded.

Increase the forces designated for peacekeeping operations Forces, ensuring that Reserves form a larger part of the mix, and include training in counter terrorism.

Cadet training programs, with their proven capability to develop good qualities of citizenship, a sense of national unity and personal pride among young Canadians, should be expanded.

NATO

The Canadian government should point out to the Europeans that Canada's presence in Europe will continue, although on a different basis and that, in any event, Canada's contributions to NATO consist of the defence of our own country (as theirs do) as well as what we do in Europe.

PEACEKEEPING

The Canadian government should press the UN Committee of 34 to find a more satisfactory way of apportioning costs for peacekeeping. A general levy on all member countries would be worth considering.

The Canadian government should press for agreement on a system of rotation of troops whenever a new mission is mounted.

ARMS CONTROL AND VERIFICATION

Arms Control negotiating and monitoring are valuable, honourable ways of contributing to a peaceful world. Canada should continue to make these important contributions.

Canada should also provide leadership for the idea of a NATO Verification Centre, as recommended by Mr. Clark in a speech delivered May 26, 1990 in Toronto.

Canada should consider a better coordination and consolidation of its maritime responsibilities, for greater effectiveness and possible savings in managing its ocean resources. Placing all non-naval responsibilities under one Minister would be one worthwhile way of approaching it.

For better coordination of assistance to civil authorities a common operations centre, similar to that of the Search and Rescue operation, should be considered.

> Given the need for a strong air surveillance capability in coastal and open waters, consideration should be given to doubling the number of Aurora or similar long range patrol aircraft from the present 18, as recommended by the Senate sub-committee on National Defence in 1983

INDUSTRIAL SUPPORT

Wherever Canada's defence industry looks for export markets, non-tariff barriers for high-technology products remain a problem. Consequently, the government should play a lead role in helping improve market access by governmental links, or by paving the way for joint ventures.

Despite the vital part a defence industrial base plays in the deterrence of war, government has not produced an overall national security policy which includes a defence industrial strategy. It is time to do so.

Because of DND's equipment requirements — both initial procurement and subsequent life-cycle support — the first priority of a defence industrial base policy should be to meet that department's needs. Therefore, DND should be designated as the lead department for producing and implementing such a policy.

The next step must be to establish a process which allows DND and industry to consult on the military's long-term requirements. These deliberations should include: what needs can be met by Canadian industry; what requirements would require a special R&D effort to be met by Canadian industry; and what items would have to be purchased offshore.

Government must devote more money and effort to research and development. Defence industries are high-technology industries and exist in a highly competitive marketplace. Research and development is their lifeblood.

Finally, no defence industrial base policy would be complete without provision for the training of new workers. The government must take the initiative in establishing apprentice-training programs to support the defence industry.

**APPENDIX A
APRIL 1991**

CANADIAN ARMED FORCES PARTICIPATION IN INTERNATIONAL PEACEKEEPING FORCES AND OBSERVER MISSIONS - 1947 ONWARDS

Location	Operation	Dates	Maximum Troop Contribution	Current Troop Contribution
Korea	UN Temporary Commission on Korea (UNTCOK)	1947-48	Unknown	—
Kashmir	UN Military Observer Group India-Pakistan (UNMOGIP)	1949-79	27	—
Korea	UN Command Korea (UNCK)	1950-54	8,000	—
Korea	UN Military Armistice Commission (UNCMAC)	1953-	2	1
Egypt Israel Jordan Lebanon Syria	UN Truce Supervisory Organization Palestine (UNTSO)	1954-	22	19
Egypt	UN Emergency Force (UNEF I)	1956-67	1,007	—
Lebanon	UN Observer Group in Lebanon (UNOGIL)	1958-59	77	—
Congo	Organisation des Nations Unies au Congo (ONUC)	1960-63	421	—
West New Guinea (now West Irian)	UN Temporary ExecutiveAuthority (UNTEA)	1962-63	13	—
Yemen	UN Yemen Observer Mission (UNYOM)	1963-64	36	—
Cyprus	UN Force in Cyprus (UNFICYP)	1964-	1,126	576

CANADIAN ARMED FORCES PARTICIPATION IN INTERNATIONAL PEACEKEEPING FORCES AND OBSERVER MISSIONS - 1947 ONWARDS (Cont'd)

Location	Operation	Dates	Maximum Troop Contribution	Current Troop Contribution
Dominican Republic	Mission of the Representatives of the Secretary-General in the Dominican Republic (DOMREP)	1965-66	1	—
India-Pakistan Border	UN India-Pakistan Observer Mission (UNIPOM)	1965-66	112	—
Egypt (Sinai)	UN Emergency Force (UNEF II)	1973-79	1,145	—
Israel Syria	UN Disengagement Observer Force (UNDOF)	1974-	220	225
Lebanon (Golan Heights)	UN Interim Force in Lebanon (UNIFIL)	1978-	117	—
Afghanistan	UN Good Offices Mission in Afghanistan and Pakistan (UNGOMAP)	1988-90	5	—
Iran/Iraq	UN Iran/Iraq Military Observer Group (UNIIMOG)	1988-91	540	—
Namibia	UN Transition Assistance Group Namibia (UNTAG)	1989-90	301	—
Central America	Grupo de Observadores de los Naçiones Uniodos en Centroamérica (ONUCA)	1989-	175	30
Afghanistan-Pakistan	Office of the Secretary-General in Afghanistan and Pakistan (OSGAP)	1990-	1	1

CANADIAN ARMED FORCES PARTICIPATION IN INTERNATIONAL PEACEKEEPING FORCES AND OBSERVER MISSIONS - 1947 ONWARDS (Cont'd)

Location	Operation	Dates	Maximum Troop Contribution	Current Troop Contribution
Haiti	UN Observers of the Election in Haiti (ONUVEH)	1990-91	11	—
Iraq-Kuwait	UN Iraq-Kuwait Observation Mission (UNIKOM)	1991-		300

OPERATIONS OUTSIDE UN FRAMEWORK

Vietnam Laos Cambodia	International Commission for Supervision and Control (ICSC)	1954-74	133	—
Nigeria	Observer Team to Nigeria (OTN)	1968-69	2	—
South Vietnam	International Commission for Control and Supervision (ICCS)	1973	248	—
Sinai	Multinational Force and Observers (MFO)	1986	140	25